21世纪高等学校计算机专业实用规划教材

Oracle Database 11g 管理与开发

杨俊生 编著

清华大学出版社
北京

内 容 简 介

在当今 IT 技术高速发展的时代，数据可用性变得越来越重要，作为主要存储技术的数据库管理系统地位越来越高。Oracle 数据库以其强大的功能和高效的性能成为数据库领域的佼佼者，成为一名优秀的 Oracle 数据库专家已成为众多 IT 精英追逐的目标。

本书基于作者多年使用 Oracle 数据库的经验和体会，全面深入地介绍了 Oracle 数据库的管理知识和开发知识：包括 Linux 环境下的数据库系统安装配置、数据库体系结构介绍、网络配置、逻辑存储结构、用户权限管理、REDO 与 UNDO 备份恢复与闪回技术、Schema 对象、SQL 开发、PL/SQL 开发以及数据库常用工具使用等内容。书中配有大量案例和深入分析，可帮助读者快速掌握 Oracle 数据库的使用。

本书可以作为高等院校计算机软件专业本科学生的 Oracle 数据库教材，也可作为 Oracle 数据库爱好者的自学读物。

本书封面贴有清华大学出版社防伪标签，无标签者不得销售。
版权所有，侵权必究。举报：010-62782989，beiqinquan@tup.tsinghua.edu.cn。

图书在版编目(CIP)数据

Oracle Database 11g 管理与开发/杨俊生编著.--北京：清华大学出版社，2014(2024.12 重印)
21 世纪高等学校计算机专业实用规划教材
ISBN 978-7-302-36150-3

Ⅰ.①O…　Ⅱ.①杨…　Ⅲ.①关系数据库系统－高等学校－教材　Ⅳ.①TP311.138

中国版本图书馆 CIP 数据核字(2014)第 072648 号

责任编辑：付弘宇　薛　阳
封面设计：何凤霞
责任校对：焦丽丽
责任印制：刘海龙

出版发行：清华大学出版社
网　　址：https://www.tup.com.cn，https://www.wqxuetang.com
地　　址：北京清华大学学研大厦 A 座　　　邮　编：100084
社 总 机：010-83470000　　　　　　　　　　邮　购：010-62786544
投稿与读者服务：010-62776969，c-service@tup.tsinghua.edu.cn
质量反馈：010-62772015，zhiliang@tup.tsinghua.edu.cn
课件下载：https://www.tup.com.cn，010-83470236

印 装 者：北京建宏印刷有限公司
经　　销：全国新华书店
开　　本：185mm×260mm　　印　张：25.5　　字　数：610 千字
版　　次：2014 年 5 月第 1 版　　　　　　　印　次：2024 年 12 月第 7 次印刷
印　　数：2761～2860
定　　价：69.80 元

产品编号：057287-02

出 版 说 明

随着我国改革开放的进一步深化,高等教育也得到了快速发展,各地高校紧密结合地方经济建设发展需要,科学运用市场调节机制,加大了使用信息科学等现代科学技术提升、改造传统学科专业的投入力度,通过教育改革合理调整和配置了教育资源,优化了传统学科专业,积极为地方经济建设输送人才,为我国经济社会的快速、健康和可持续发展以及高等教育自身的改革发展做出了巨大贡献。但是,高等教育质量还需要进一步提高以适应经济社会发展的需要,不少高校的专业设置和结构不尽合理,教师队伍整体素质亟待提高,人才培养模式、教学内容和方法需要进一步转变,学生的实践能力和创新精神亟待加强。

教育部一直十分重视高等教育质量工作。2007 年 1 月,教育部下发了《关于实施高等学校本科教学质量与教学改革工程的意见》,计划实施"高等学校本科教学质量与教学改革工程(简称'质量工程')",通过专业结构调整、课程教材建设、实践教学改革、教学团队建设等多项内容,进一步深化高等学校教学改革,提高人才培养的能力和水平,更好地满足经济社会发展对高素质人才的需要。在贯彻和落实教育部"质量工程"的过程中,各地高校发挥师资力量强、办学经验丰富、教学资源充裕等优势,对其特色专业及特色课程(群)加以规划、整理和总结,更新教学内容、改革课程体系,建设了一大批内容新、体系新、方法新、手段新的特色课程。在此基础上,经教育部相关教学指导委员会专家的指导和建议,清华大学出版社在多个领域精选各高校的特色课程,分别规划出版系列教材,以配合"质量工程"的实施,满足各高校教学质量和教学改革的需要。

本系列教材立足于计算机专业课程领域,以专业基础课为主、专业课为辅,横向满足高校多层次教学的需要。在规划过程中体现了如下一些基本原则和特点。

(1) 反映计算机学科的最新发展,总结近年来计算机专业教学的最新成果。内容先进,充分吸收国外先进成果和理念。

(2) 反映教学需要,促进教学发展。教材要适应多样化的教学需要,正确把握教学内容和课程体系的改革方向,融合先进的教学思想、方法和手段,体现科学性、先进性和系统性,强调对学生实践能力的培养,为学生知识、能力、素质协调发展创造条件。

(3) 实施精品战略,突出重点,保证质量。规划教材把重点放在公共基础课和专业基础课的教材建设上;特别注意选择并安排一部分原来基础比较好的优秀教材或讲义修订再版,逐步形成精品教材;提倡并鼓励编写体现教学质量和教学改革成果的教材。

(4) 主张一纲多本,合理配套。专业基础课和专业课教材配套,同一门课程有针对不同层次、面向不同应用的多本具有各自内容特点的教材。处理好教材统一性与多样化,基本教材与辅助教材、教学参考书,文字教材与软件教材的关系,实现教材系列资源配套。

(5) 依靠专家,择优选用。在制定教材规划时要依靠各课程专家在调查研究本课程教材建设现状的基础上提出规划选题。在落实主编人选时，要引入竞争机制，通过申报、评审确定主题。书稿完成后要认真实行审稿程序，确保出书质量。

　　繁荣教材出版事业，提高教材质量的关键是教师。建立一支高水平教材编写梯队才能保证教材的编写质量和建设力度，希望有志于教材建设的教师能够加入到我们的编写队伍中来。

21世纪高等学校计算机专业实用规划教材

联系人：魏江江　weijj@tup.tsinghua.edu.cn

前　言

　　计算机软件技术经过多年发展,已经应用于各行各业,数据可用性直接关系到企业的生存与发展,而数据库管理系统作为一种有效的数据存储方式已经变得越来越重要。Oracle 数据库是当前主流的关系数据库管理系统之一,市场占有率遥遥领先。Oracle 11g 是当前最流行的 Oracle 数据库版本,支持多种操作系统平台,广泛应用于通信、金融、政府等行业。

　　Oracle 数据库的市场高占有率以及强大的功能吸引了无数 IT 从业者的眼球,成为一名精通 Oracle 数据库的 DBA 逐渐成了很多人从业的终极目标。要想精通 Oracle 数据库并非易事,无数 IT 精英经历着学习 Oracle 的困惑,从初学的懵懂,到拨开云雾见天日的豁然开朗,再到夜以继日的深入研究,最后变成了对 Oracle 数据库技术的痴迷。要想学好 Oracle 数据库,单纯学习数据库本身的技术是不够的,需要一些综合知识,至少需要掌握一种 Linux 或 UNIX 系统,懂些存储方面的知识,掌握一门开发语言,懂得当前流行的软件体系结构,还要懂些与网络相关的知识。

　　本书结合作者多年从事 Oracle 数据库管理系统使用经验,以及多年从事 Oracle 数据库培训和教学经验,从 Oracle 数据库的实际使用出发,系统地介绍了 Oracle 11g 数据库的管理和开发。全书分为 4 个部分,共 13 章。

　　第一部分:Oracle 数据库安装配置,包括第 1 章。详细讲解了 Linux 环境下 Oracle 数据库的安装配置,给出详细的配置步骤,供学习者自行安装时参考。

　　第二部分:Oracle 数据库日常管理,包括 6 章,从第 2 章到第 7 章。第 2 章讲解 Oracle 数据库体系结构,以及数据库和实例各组成部分;第 3 章讲解 Oracle 数据库的网络环境配置;第 4 章讲解 Oracle 数据库逻辑结构,以及表空间、段、区和块之间的关系;第 5 章讲解用户权限管理;第 6 章讲解 REDO 和 UNDO 的作用及配置;第 7 章讲解 Oracle 数据库备份和恢复知识。

　　第三部分:Oracle 数据库开发,包括 5 章,从第 8 章到第 12 章。第 8 章讲解 Oracle 数据库 Schema 对象,重点讲解各个对象的应用场景;第 9 章讲解 SQL 语句,重点讲解 DML 语句和 SELECT 语句;第 10 章讲解 Oracle 的内置函数;第 11 章讲解 PL/SQL 开发基础知识;第 12 章讲解 PL/SQL 程序设计开发。

　　第四部分:Oracle 数据库工具使用,包括第 13 章。讲解 Oracle 常用工具的使用,包括 SQL * Plus、SQL * Loader 以及 EXPDP 和 IMPDP。

　　在本书的写作过程中得到了清华大学出版社的大力支持,特此表示衷心感谢,同时非常感谢家人的默默支持。

    鉴于作者才学有限，书中难免存在不足之处，欢迎同行、专家及读者批评指正。联系邮箱：yangjs@dlut.edu.cn，QQ号：908464122。

编 者

2014 年 2 月

目 录

第一部分　Oracle 数据库安装配置

第 1 章　安装配置 3

- 1.1　下载 Oracle 11g 安装包 3
- 1.2　安装 Oracle 11g 软件 4
  - 1.2.1　检查硬件要求 4
  - 1.2.2　检查软件要求 5
  - 1.2.3　安装 Linux 操作系统 6
  - 1.2.4　配置 Oracle 11g 安装环境 19
  - 1.2.5　安装 Oracle 11g 数据库软件 21
- 1.3　配置数据库监听器 29
  - 1.3.1　使用 NETCA 配置监听器 29
  - 1.3.2　使用 LSNRCTL 启动监听器 32
- 1.4　创建数据库 33
- 1.5　配置客户端 43
  - 1.5.1　安装 Oracle 11g 客户端软件 43
  - 1.5.2　使用 NETCA 配置网络服务命名 47
  - 1.5.3　连接数据库 51
- 小结 53
- 思考题 53

第二部分　Oracle 数据库日常管理

第 2 章　数据库体系结构 57

- 2.1　数据库体系结构简介 57
- 2.2　数据库实例结构 58
  - 2.2.1　数据库实例简介 58
  - 2.2.2　实例内存结构 60
  - 2.2.3　Oracle 实例进程结构 66
- 2.3　数据库物理结构 67

2.3.1　数据文件和临时文件 …………………………………… 68
　　2.3.2　控制文件 …………………………………………………… 68
　　2.3.3　在线日志文件 ……………………………………………… 69
2.4　参数文件管理 ……………………………………………………… 69
　　2.4.1　参数文件介绍和管理 ……………………………………… 69
　　2.4.2　基本参数介绍 ……………………………………………… 71
2.5　口令文件管理 ……………………………………………………… 72
2.6　数据库启动和停止 ………………………………………………… 74
　　2.6.1　数据库启动 ………………………………………………… 74
　　2.6.2　数据库停止 ………………………………………………… 78
小结 ………………………………………………………………………… 80
思考题 ……………………………………………………………………… 80

第 3 章　网络配置 …………………………………………………………… 82

3.1　网络服务组件 ……………………………………………………… 82
　　3.1.1　Oracle Net ………………………………………………… 82
　　3.1.2　Oracle Net Listener ……………………………………… 83
3.2　数据库服务注册 …………………………………………………… 84
　　3.2.1　服务的概念 ………………………………………………… 85
　　3.2.2　服务注册 …………………………………………………… 86
3.3　监听器配置 ………………………………………………………… 87
　　3.3.1　监听器配置文件 …………………………………………… 87
　　3.3.2　监听器配置与管理 ………………………………………… 89
3.4　客户端连接 ………………………………………………………… 95
　　3.4.1　配置本地命名解析 ………………………………………… 95
　　3.4.2　本地命名解析文件 ………………………………………… 99
小结 ………………………………………………………………………… 101
思考题 ……………………………………………………………………… 101

第 4 章　逻辑存储结构 ……………………………………………………… 102

4.1　逻辑结构 …………………………………………………………… 102
　　4.1.1　逻辑结构简介 ……………………………………………… 102
　　4.1.2　区管理方式 ………………………………………………… 103
　　4.1.3　段空间管理方式 …………………………………………… 104
4.2　数据块 ……………………………………………………………… 107
　　4.2.1　数据块结构 ………………………………………………… 107
　　4.2.2　行数据存储格式 …………………………………………… 109
4.3　区 …………………………………………………………………… 110
4.4　段 …………………………………………………………………… 111

    4.4.1 用户段创建 ····················· 112
    4.4.2 临时段和 undo 段 ················ 112
  4.5 表空间 ·························· 113
    4.5.1 表空间介绍 ····················· 113
    4.5.2 表空间基本操作 ················ 114
  4.6 相关视图 ························ 119
  小结 ······························· 121
  思考题 ···························· 121

第 5 章 用户权限管理 ···················· 123

  5.1 用户管理 ························ 123
    5.1.1 创建用户 ······················· 123
    5.1.2 修改用户 ······················· 127
    5.1.3 概要文件 ······················· 129
    5.1.4 删除用户 ······················· 132
    5.1.5 相关视图 ······················· 132
  5.2 权限管理 ························ 132
    5.2.1 系统权限 ······················· 133
    5.2.2 对象权限 ······················· 135
    5.2.3 相关视图 ······················· 137
  5.3 角色管理 ························ 138
    5.3.1 自定义角色 ···················· 138
    5.3.2 角色授予与回收 ················ 139
    5.3.3 相关视图 ······················· 140
  5.4 安全审计 ························ 141
    5.4.1 标准审计 ······················· 141
    5.4.2 精度审计 ······················· 146
    5.4.3 相关视图 ······················· 147
  小结 ······························· 149
  思考题 ···························· 149

第 6 章 REDO 与 UNDO ···················· 150

  6.1 REDO ···························· 150
    6.1.1 REDO 概述 ······················ 150
    6.1.2 日志组 ·························· 151
    6.1.3 日志维护 ······················· 152
    6.1.4 相关视图 ······················· 156
  6.2 UNDO ···························· 156
    6.2.1 UNDO 概述 ····················· 156

6.2.2　UNDO 段空间使用 ································ 158
　　　6.2.3　UNDO 管理 ····································· 159
　　　6.2.4　相关视图 ······································· 160
　6.3　检查点 ··· 162
小结 ·· 164
思考题 ·· 164

第 7 章　备份恢复与闪回技术 ································ 165

　7.1　备份恢复概述 ··· 165
　　　7.1.1　备份文件 ······································· 165
　　　7.1.2　数据库备份 ····································· 166
　　　7.1.3　数据库恢复 ····································· 167
　7.2　恢复管理器 ··· 171
　　　7.2.1　恢复管理器介绍 ································· 171
　　　7.2.2　RMAN 命令介绍 ································· 173
　7.3　闪回技术 ··· 184
　　　7.3.1　闪回查询 ······································· 184
　　　7.3.2　闪回数据 ······································· 185
　　　7.3.3　闪回删除 ······································· 186
　　　7.3.4　闪回版本查询 ··································· 187
　　　7.3.5　闪回事务 ······································· 188
　　　7.3.6　闪回数据库 ····································· 188
　　　7.3.7　闪回归档 ······································· 189
小结 ·· 190
思考题 ·· 190

第三部分　Oracle 数据库开发

第 8 章　Schema 对象 ······································ 193

　8.1　表 ··· 193
　　　8.1.1　数据类型 ······································· 193
　　　8.1.2　堆表 ··· 196
　　　8.1.3　临时表 ··· 199
　　　8.1.4　索引组织表 ····································· 200
　　　8.1.5　分区表 ··· 201
　　　8.1.6　相关视图 ······································· 210
　8.2　视图 ··· 211
　　　8.2.1　普通视图 ······································· 211
　　　8.2.2　物化视图 ······································· 213

		8.2.3　相关视图	216
8.3	索引		217
		8.3.1　索引类别	217
		8.3.2　索引维护	219
		8.3.3　相关视图	221
8.4	簇		221
8.5	序列		223
8.6	同义词		225
8.7	数据库链接		226
8.8	约束		227
		8.8.1　约束分类	227
		8.8.2　约束操作	228
		8.8.3　相关视图	232

小结 233

思考题 233

第 9 章　SQL 开发　234

9.1	结构化查询语言(SQL)简介	234
9.2	数据操纵语句	235
	9.2.1　INSERT 语句	235
	9.2.2　UPDATE 语句	239
	9.2.3　DELETE 语句	240
	9.2.4　MERGE 语句	241
9.3	查询语句	242
	9.3.1　单表查询	242
	9.3.2　连接查询	250
	9.3.3　子查询	257
	9.3.4　高级查询	262
9.4	事务控制语句	269
	9.4.1　Oracle 事务开始结束条件	269
	9.4.2　事务控制语句	270

小结 271

思考题 271

第 10 章　内置函数　273

10.1	内置函数简介	273
10.2	字符函数	273
10.3	数字函数	277
10.4	日期函数	279

10.5 转换函数 ………………………………………………………………………………… 282
10.6 正则表达式函数 …………………………………………………………………………… 285
小结 …………………………………………………………………………………………… 287
思考题 ………………………………………………………………………………………… 287

第 11 章 PL/SQL 基础 …………………………………………………………………… 288

11.1 PL/SQL 简介 ……………………………………………………………………………… 288
11.2 基本块结构 ………………………………………………………………………………… 288
11.3 变量类型 …………………………………………………………………………………… 290
  11.3.1 标量数据类型 …………………………………………………………………… 290
  11.3.2 复合数据类型 …………………………………………………………………… 292
  11.3.3 集合运算符 ……………………………………………………………………… 297
  11.3.4 集合方法 ………………………………………………………………………… 298
11.4 变量声明赋值 ……………………………………………………………………………… 299
11.5 控制结构 …………………………………………………………………………………… 300
  11.5.1 条件语句 ………………………………………………………………………… 300
  11.5.2 循环语句 ………………………………………………………………………… 302
11.6 游标 ………………………………………………………………………………………… 306
  11.6.1 隐式游标 ………………………………………………………………………… 306
  11.6.2 显式游标 ………………………………………………………………………… 307
11.7 批语句 ……………………………………………………………………………………… 310
  11.7.1 BULK COLLECT INTO 语句 ……………………………………………… 310
  11.7.2 FORALL 语句 …………………………………………………………………… 312
11.8 异常处理 …………………………………………………………………………………… 313
  11.8.1 预定义异常 ……………………………………………………………………… 314
  11.8.2 用户定义异常 …………………………………………………………………… 314
11.9 动态 SQL …………………………………………………………………………………… 316
小结 …………………………………………………………………………………………… 317
思考题 ………………………………………………………………………………………… 318

第 12 章 PL/SQL 程序设计 ……………………………………………………………… 319

12.1 函数 ………………………………………………………………………………………… 319
12.2 过程 ………………………………………………………………………………………… 324
12.3 调用过程和函数 …………………………………………………………………………… 325
12.4 包 …………………………………………………………………………………………… 326
  12.4.1 包规范 …………………………………………………………………………… 326
  12.4.2 包体 ……………………………………………………………………………… 328
12.5 触发器 ……………………………………………………………………………………… 329
  12.5.1 数据定义触发器 ………………………………………………………………… 330

12.5.2　数据操作触发器 ……………………………………………… 332
　　12.5.3　复合触发器 …………………………………………………… 334
　　12.5.4　INSTEAD-OF 触发器 …………………………………………… 335
　　12.5.5　系统事件触发器 ………………………………………………… 336
　　12.5.6　触发器编译和启用 ……………………………………………… 337
12.6　相关视图 ……………………………………………………………… 337
小结 …………………………………………………………………………… 338
思考题 ………………………………………………………………………… 338

第四部分　Oracle 数据库工具使用

第 13 章　工具使用 ………………………………………………………… 343

13.1　SQL * Plus ……………………………………………………………… 343
　　13.1.1　启动 SQL * Plus ………………………………………………… 343
　　13.1.2　SQL * Plus 环境变量设置 ……………………………………… 344
　　13.1.3　SQL * Plus 配置 ………………………………………………… 346
　　13.1.4　SQL * Plus 连接数据库 ………………………………………… 347
　　13.1.5　SQL * Plus 命令 ………………………………………………… 350
13.2　SQL * Loader ………………………………………………………… 359
　　13.2.1　启动 SQL * Loader ……………………………………………… 360
　　13.2.2　SQL * Loader 命令行参数 ……………………………………… 361
　　13.2.3　控制文件格式说明 ……………………………………………… 366
　　13.2.4　指定加载数据文件 ……………………………………………… 367
　　13.2.5　指定数据分隔方式 ……………………………………………… 368
　　13.2.6　指定条件 ………………………………………………………… 369
　　13.2.7　多表导入 ………………………………………………………… 370
　　13.2.8　指定列及数据类型 ……………………………………………… 371
13.3　EXPDP ………………………………………………………………… 373
　　13.3.1　启动 EXPDP ……………………………………………………… 374
　　13.3.2　EXPDP 导出模式 ………………………………………………… 375
　　13.3.3　EXPDP 命令行参数 ……………………………………………… 375
　　13.3.4　交互模式 ………………………………………………………… 381
13.4　IMPDP ………………………………………………………………… 383
　　13.4.1　启动 IMPDP ……………………………………………………… 383
　　13.4.2　IMPDP 导入模式 ………………………………………………… 384
　　13.4.3　IMPDP 命令行参数 ……………………………………………… 385
　　13.4.4　交互模式 ………………………………………………………… 389
小结 …………………………………………………………………………… 390
思考题 ………………………………………………………………………… 391

第一部分

Oracle数据库安装配置

第 1 章　　安 装 配 置

生产环境中 Oracle 数据库通常安装在 UNIX 系统或 Linux 系统中，而初学者往往在安装阶段弄得焦头烂额。本章将详细介绍 Red Hat Enterprise Linux 环境下的 Oracle 11g 安装配置。

1.1　下载 Oracle 11g 安装包

登录 Oracle 的官方网站 www.oracle.com，免费注册账号并登录，然后选择 Downloads | Oracle Database 选项，进入 Oracle 数据库的下载界面，如图 1-1 所示。

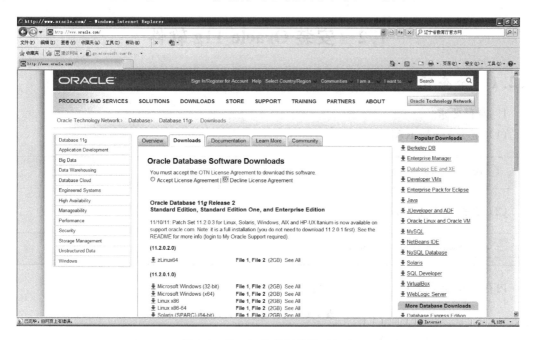

图 1-1　Oracle 软件下载界面（一）

选择 Accept License Agreement 后的界面如图 1-2 所示。

接下来选择相应平台进行下载，本书中使用的下载版本为 Linux X86 平台对应的版本，下载文件为两个压缩包，占磁盘空间 2GB 左右。

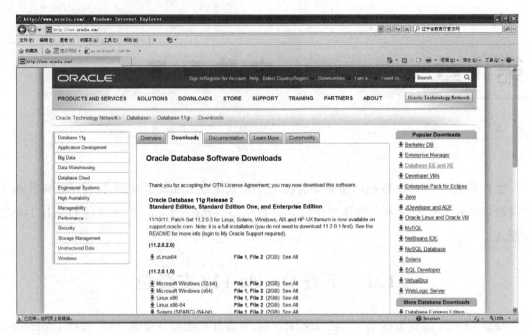

图 1-2　Oracle 软件下载界面(二)

1.2　安装 Oracle 11g 软件

1.2.1　检查硬件要求

1. 内存要求

(1) 至少 256M RAM。

```
# grep MemTotal /proc/meminfo
```

(2) 交换区大小建议。

```
# grep SwapTotal /proc/meminfo
```

内存与交换区配置如表 1-1 所示。

表 1-1　内存与交换区配置

可用 RAM	交　换　区
0～256MB	RAM 3 倍大小
256～512MB	RAM 2 倍大小
512MB～2GB	RAM 1.5 倍大小
2～16GB	RAM 1 倍大小
16GB 以上	16GB

建议物理内存不低于 2GB，如果使用虚拟机，虚拟机的内存不低于 1GB。

2. 磁盘空间要求

不同版本对磁盘空间的要求如表 1-2 所示。

表 1-2　不同版本的空间需求

安装类型	软件空间需求	初始数据库空间需求
企业版 32 位	3.95GB	1.7GB
标准版 32 位	3.88GB	1.5GB

建议可用空间 10GB 以上。

1.2.2　检查软件要求

1. 操作系统要求

Oracle 11g Release 2 (11.2)对操作系统版本要求：

(1) Asianux Server 3 SP2

(2) Oracle Linux 4 Update 7

(3) Oracle Linux 5 Update 2

(4) Red Hat Enterprise Linux 4 Update 7

(5) Red Hat Enterprise Linux 5 Update 2

(6) SUSE Linux Enterprise Server 10 SP2

(7) SUSE Linux Enterprise Server 11

本书选择 Red Hat Enterprise Linux 5，通过如下命令查询：

```
# cat /proc/version
```

2. 内核要求

Oracle 11g Release 2 (11.2)对操作系统内核要求：

(1) Oracle Linux 4 and Red Hat Enterprise Linux 4：2.6.9 或更新版本。

(2) Asianux Server 3, Oracle Linux 5, and Red Hat Enterprise Linux 5：2.6.18 或更新版本。

(3) SUSE Linux Enterprise Server 10：2.6.16.21 或更新版本。

(4) SUSE Linux Enterprise Server 11：2.6.27.19 或更新版本。

版本信息通过如下命令查询：

```
# uname -r
```

3. 依赖包要求

对于 Red Hat Enterprise Linux 5，要求安装包如下：

binutils-2.17.50.0.6

compat-libstdc++-33-3.2.3

elfutils-libelf-0.125
elfutils-libelf-devel-0.125
elfutils-libelf-devel-static-0.125
gcc-4.1.2
gcc-c++-4.1.2
glibc-2.5-24
glibc-common-2.5
glibc-devel-2.5
ksh-20060214
libaio-0.3.106
libaio-devel-0.3.106
libgcc-4.1.2
libgomp-4.1.2
ibstdc++-4.1.2
libstdc++-devel-4.1.2
make-3.81
sysstat-7.0.2

1.2.3 安装 Linux 操作系统

在 Windows XP 系统中可通过使用虚拟机软件安装 Linux 系统，首先安装虚拟机软件，然后创建虚拟机，之后再安装 Linux 系统。本书使用的虚拟机软件是 VMware Server，虚拟机软件安装比较简单，此处略过，接下来讲解虚拟机创建和 Linux 系统安装步骤。

1. 创建虚拟机

打开虚拟机软件，然后选择 File|New|Virtual Machine，显示虚拟机创建页面。在如图 1-3 所示的页面中选择 Custom 单选按钮，单击【下一步】按钮。在如图 1-4 所示页面中选择操作系统 Linux 和 Red Hat Linux，单击【下一步】按钮。

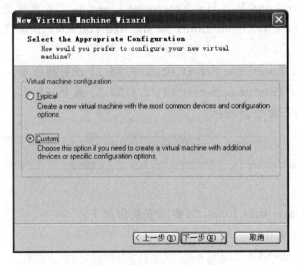

图 1-3 虚拟机配置类型

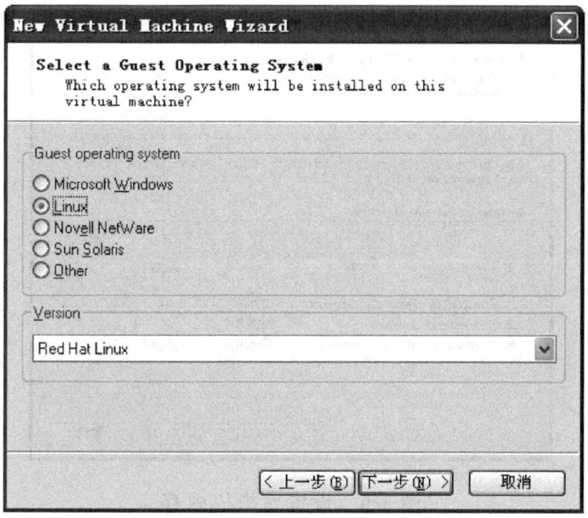

图 1-4　虚拟机操作系统

在如图 1-5 所示的页面中输入虚拟机名称 Oracle 11g，选择虚拟机创建路径 C:\Linux，单击【下一步】按钮，接下来几个页面保持默认选项即可。在如图 1-6 所示页面中设置虚拟机使用的内存，可根据物理内存大小合理设置，此处为 1024MB，建议不低于 1024MB，单击【下一步】按钮。

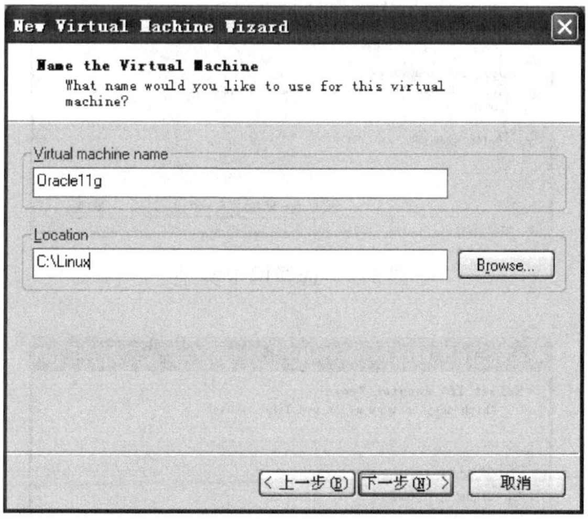

图 1-5　设置虚拟机名称和路径

在如图 1-7 所示的页面中设置虚拟网卡类型，此处选择 Use host-only networking 单选按钮，单击【下一步】按钮。在如图 1-8 所示的页面选择 IO 适配器类型 SCSI Adapters|LSI Logic，单击【下一步】按钮。

在如图 1-9 所示的页面中选择 Create a new virtual disk 单选按钮，单击【下一步】按钮。在如图 1-10 所示的页面中选择虚拟磁盘类型 SCSI，单击【下一步】按钮。

在如图 1-11 所示的页面中设置虚拟磁盘大小，此处设为 20GB，选择 Allocate all disk space now 复选框，单击【下一步】按钮，开始创建磁盘。磁盘创建完成后虚拟机就已经准备好了。

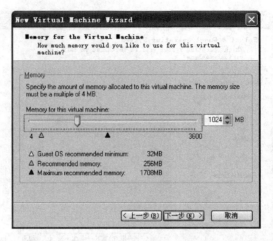

图 1-6 设置虚拟机内存

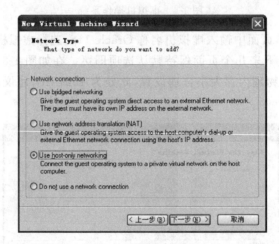

图 1-7 选择网卡类型

图 1-8 选择 IO 适配器类型

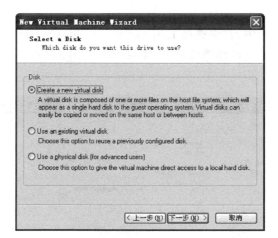

图 1-9 选择磁盘

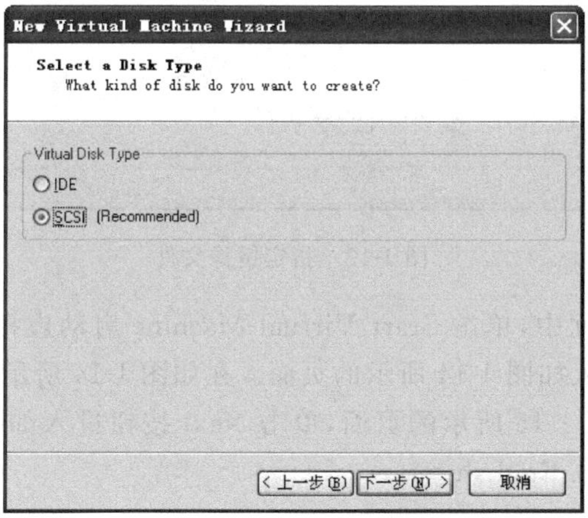

图 1-10 选择磁盘类型

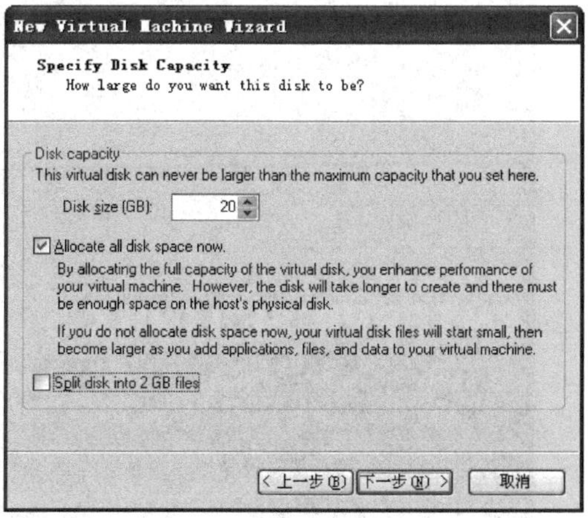

图 1-11 磁盘大小设置页面

2. 安装 Linux 系统

在虚拟机软件主页面中，首先选择 Edit virtual machine settings，在如图 1-12 所示页面中指定安装 Linux 操作系统的镜像文件，在 Use ISO image 处选择镜像文件路径，然后单击 OK 按钮关闭对话框。

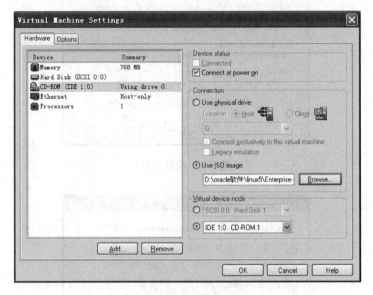

图 1-12　指定镜像文件

在虚拟机软件主页面中，单击 Start Virtual Machine 启动虚拟机，进入如图 1-13 所示的页面，按 Enter 键，进入如图 1-14 所示的页面。在如图 1-14 所示的页面中选择 Skip 跳过安装介质检查，进入如图 1-15 所示的页面，单击 Next 按钮进入如图 1-16 所示的页面进行语言选择，默认选择 U. S. English。

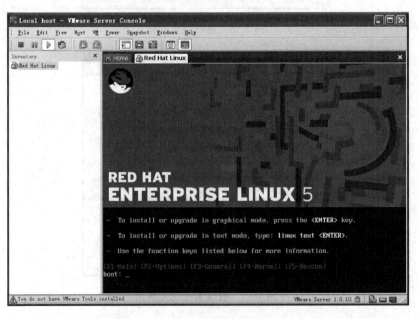

图 1-13　安装类型选择

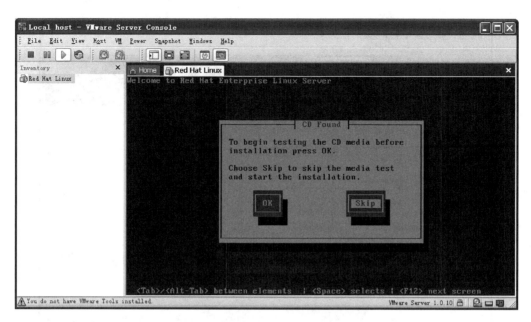

图 1-14　安装介质检查页面

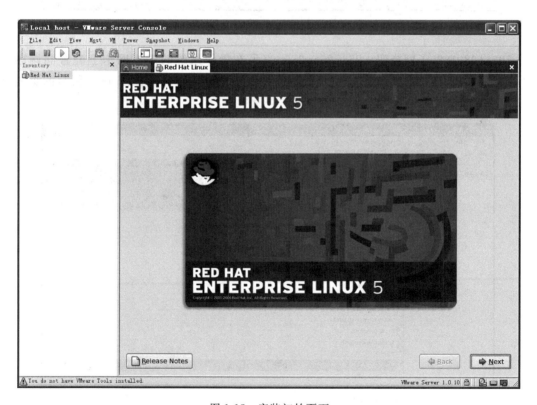

图 1-15　安装初始页面

在如图 1-16 所示的页面中单击 Next 按钮进入如图 1-17 所示的页面进行键盘类型选择，默认选择 U.S.English，单击 Next 按钮，进入如图 1-18 所示的页面，单击 Edit 按钮输入

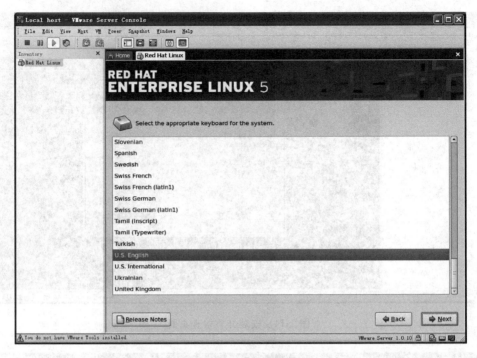

图 1-16　语言选择页面

IP：192.168.1.2，子网掩码：255.255.255.0（另在操作系统的虚拟网卡上配置同网段的 IP：192.168.1.1）。在如图 1-18 所示的页面中单击 OK 按钮进入如图 1-19 所示的页面设置主机名：oracle11g，单击 Next 按钮。

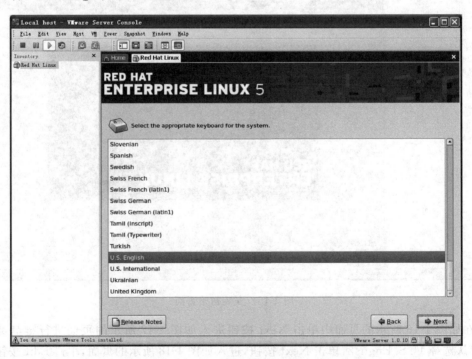

图 1-17　键盘选择页面

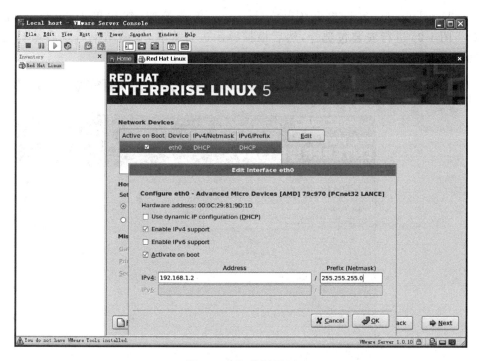

图 1-18　IP 设置页面

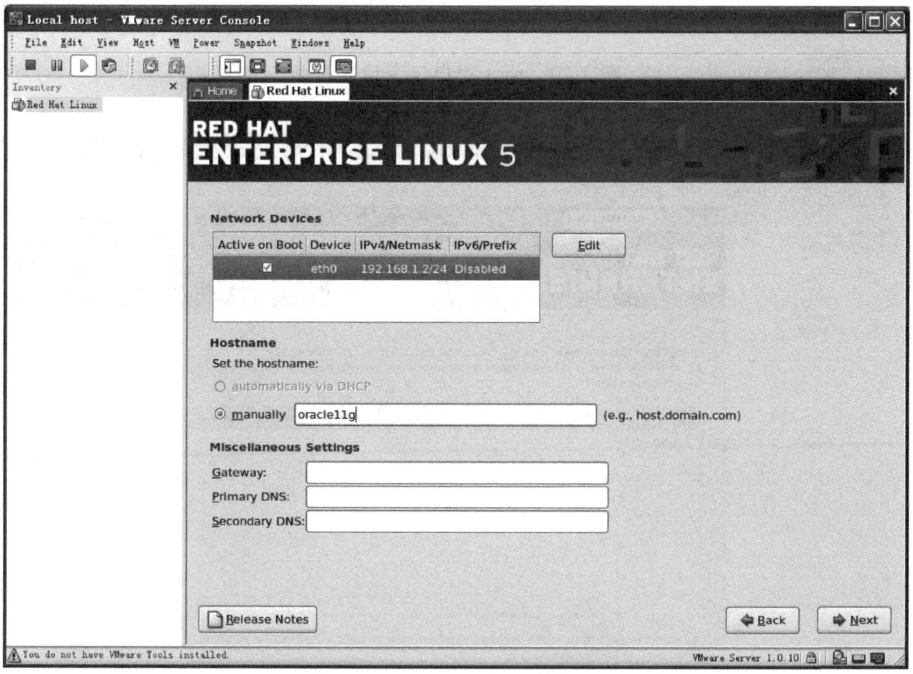

图 1-19　主机名设置页面

在如图 1-20 所示的页面中设置 root 用户口令，然后单击 Next 按钮进入如图 1-21 所示的页面进行安装包定制（或安装系统后使用 Linux 系统命令 yum 来安装所需的包），选中

Software Development 复选框和 Customize now 单选按钮,然后单击 Next 按钮进入图 1-22～图 1-25 进行安装包选择,然后根据提示进行安装,安装结束后重启系统进入如图 1-26 所示的页面进行系统设置。

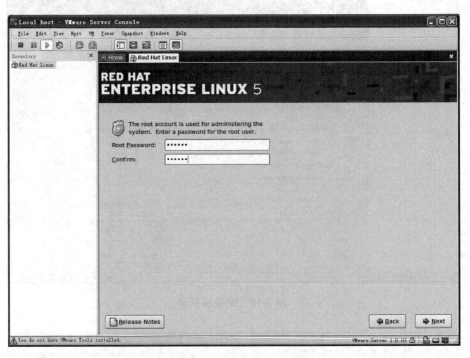

图 1-20　root 用户口令设置页面

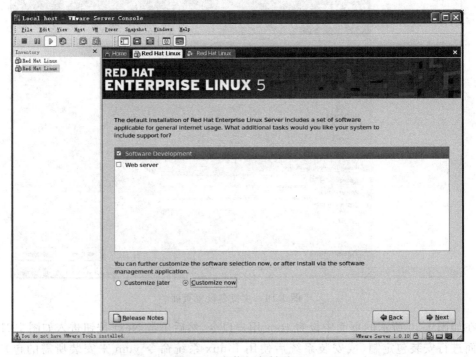

图 1-21　安装包定制页面

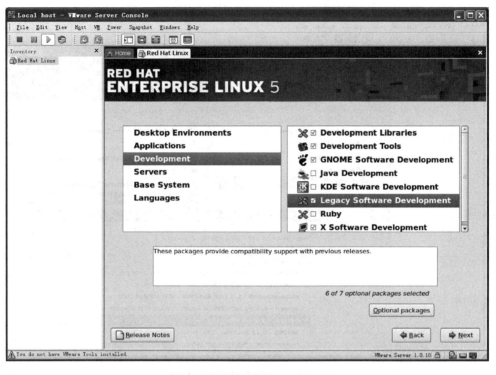

图 1-22 安装包选择页面(一)

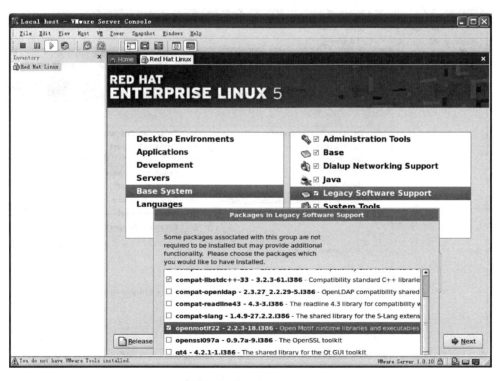

图 1-23 安装包选择页面(二)

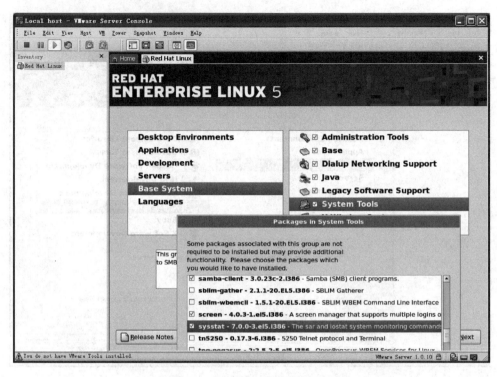

图 1-24　安装包选择页面(三)

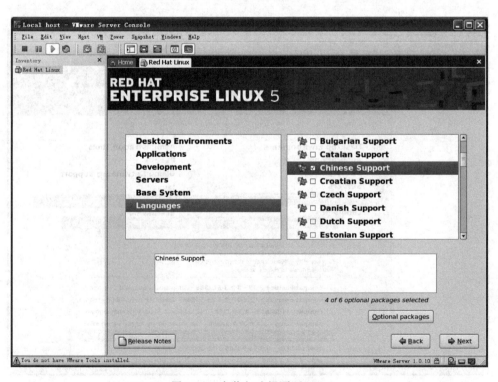

图 1-25　安装包选择页面(四)

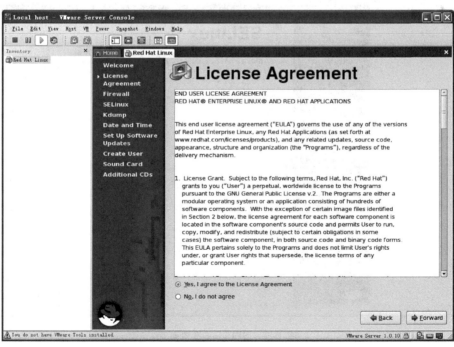

图 1-26　系统重启后欢迎页面

按照提示分别在如图 1-27 所示的页面中单击 Yes,I agree to the License Agreement 单选按钮,在如图 1-28 和图 1-29 所示的页面中选择 Disabled,接下来根据提示进行设置,最后重启系统,安装结束。

图 1-27　接受协议页面

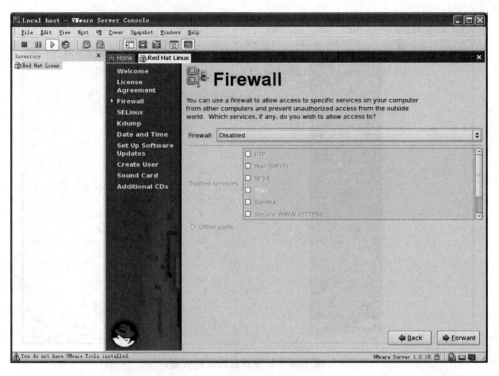

图 1-28　防火墙设置页面

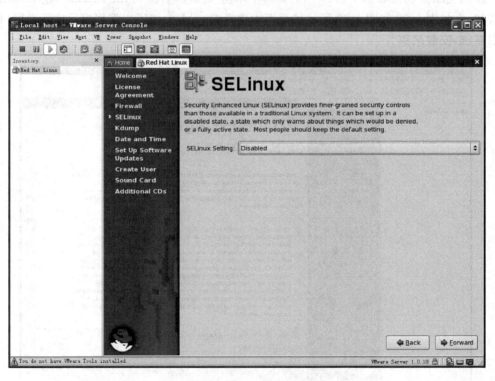

图 1-29　SELinux 设置页面

1.2.4 配置 Oracle 11g 安装环境

1. 系统内核参数设置

Oracle 数据库软件安装对操作系统内核参数都有基本要求,具体参数值参考 Oracle 11g 的安装手册。编辑/etc/sysctl.conf 文件,添加如下内容到文件尾:

```
fs.aio-max-nr = 1048576
fs.file-max = 6815744
kernel.shmall = 2097152
kernel.shmmax = 536870912
kernel.shmmni = 4096
kernel.sem = 250 32000 100 128
net.ipv4.ip_local_port_range = 9000 65500
net.core.rmem_default = 262144
net.core.rmem_max = 4194304
net.core.wmem_default = 262144
net.core.wmem_max = 1048586
```

输入如下命令使参数修改生效:

```
# /sbin/sysctl -p
```

输入如下命令查看系统参数:

```
# /etc/sysctl -a | grep 参数名
```

2. 用户和组设置

安装 Oracle 数据库所需用户和组如表 1-3 所示。

表 1-3 安装 Oracle 数据库所需用户和组

类型	名称
用户组	oinstall
用户组	dba
用户组	oper(可选)
用户	oracle

添加组 oinstall 和 dba:

```
# groupadd oinstall
# groupadd dba
```

添加用户 oracle:

```
# useradd -g oinstall -G dba oracle
```

设置用户口令：

```
# passwd oracle
```

3. 资源限制

Oracle 建议对账户可以使用的进程数和打开的文件数进行设置，添加如下内容至文件 /etc/security/limits.conf 末尾：

```
oracle    soft    nproc     2047
oracle    hard    nproc     16384
oracle    soft    nofile    1024
oracle    hard    nofile    65536
oracle    soft    stack     10240
```

limits.conf 文件内容修改后可立即生效，重新以 oracle 用户登录，验证设置结果。
检查文件描述符数量限制：

```
$ ulimit  -Sn
$ ulimit  -Hn
```

检查进程数量限制：

```
$ ulimit  -Su
$ ulimit  -Hu
```

检查进程栈大小限制：

```
$ ulimit  -Ss
$ ulimit  -Hs
```

4. 目录设置

创建一个 Oracle 数据库软件安装目录，并进行目录所有者和访问权限设置：

```
# mkdir  -p  /u01/app/product/11.2.0/dbhome_1
# chown  -R  oracle:oinstall /u01
# chmod  -R  755 /u01
```

5. 用户环境变量设置

以 oracle 用户登录，编辑 $HOME/.bash_profile，将如下内容添加至文件末尾：

```
export ORACLE_BASE = /u01/app
export ORACLE_HOME = $ORACLE_BASE/product/11.2.0/dbhome_1
export ORACLE_SID = mysid
export PATH = $ORACLE_HOME/bin:$PATH
export EDITOR = vi
```

1.2.5 安装 Oracle 11g 数据库软件

通过 FTP 上传并解压 Oracle 11g 软件安装包,然后进入解压目录执行:

```
$ ./runInstaller
```

命令执行后,弹出如图 1-30 所示的安装界面,在页面中可以设置接收安全问题的 Email 以及账号口令(此处忽略),单击 Next 按钮。图 1-31 中显示出安装选项。选项一 Create and configure a database,安装软件后直接创建数据库;选项二 Install database software only,只安装数据库软件;选项三 Upgrade an existing database,升级存在的数据库到新版本。此处选择 Install database software only 单选按钮。

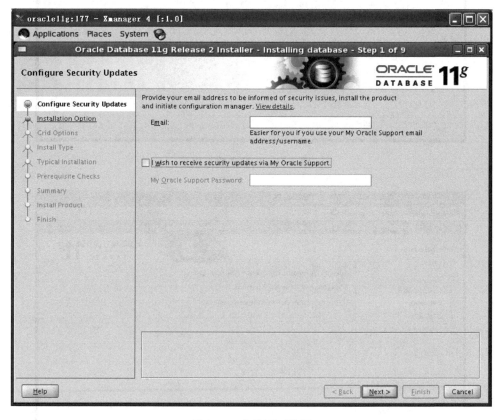

图 1-30　安装页面首页

图 1-32 中显示数据库的类型。类型一 Single instance database installation,单实例数据库;类型二 Real Application Clusters database installation,集群数据库 RAC。此处选择类型 Single instance database installation 单选按钮,即单实例数据库。

图 1-33 中显示了产品支持语言,此处默认选择 English。

图 1-34 显示了数据库版本。版本一 Enterprise Edition,企业版本;版本二 Standard Edition,标准版本;版本三 Standard Edition One,标准版本 1。此处选择 Enterprise Edition 单选按钮,即企业版。

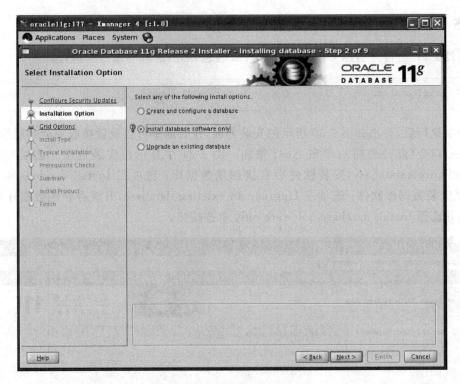

图 1-31　安装选项页面

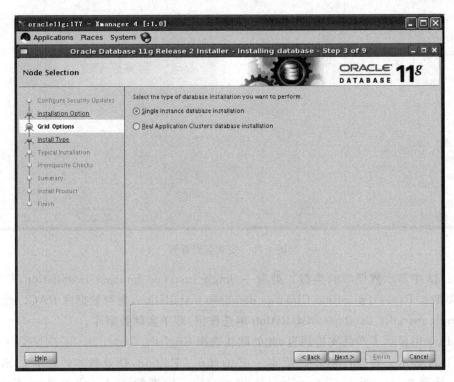

图 1-32　数据库类型选择界面

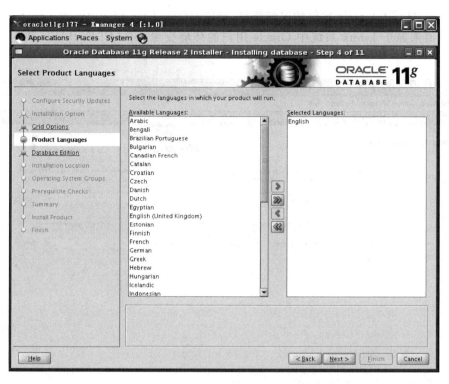

图 1-33　产品语言选择页面

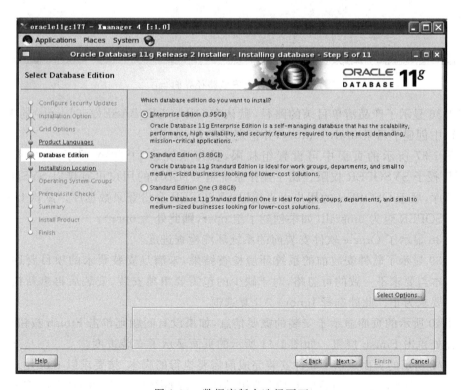

图 1-34　数据库版本选择页面

图 1-35 显示了 Oracle 软件的安装路径,指定 Oracle Base 和 Software Location,这两个输入框的值来源于 Linux 系统中 Oracle 用户的环境变量 ORACLE_BASE 和 ORACLE_HOME。

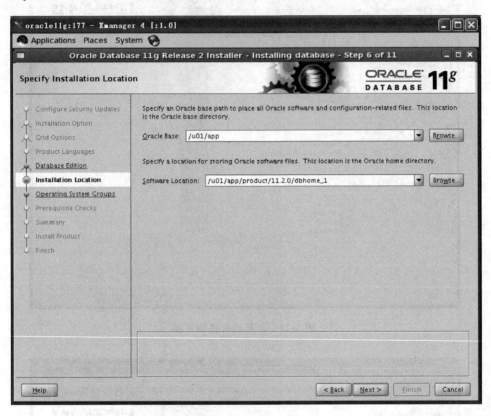

图 1-35　安装路径界面

图 1-36 显示了产品清单目录的路径,默认在 ORACLE_BASE(/u01/app)的上层目录中(/u01)中创建子目录 oraInventory,并指定了组 oinstall。

在图 1-37 所示的页面中可设置组权限,操作系统组 OSDBA 授予 SYSDBA 权限,OSOPER 授予 SYSOPER 权限。属于操作系统组 OSDBA 的用户可直接以 SYSDBA 身份登录数据库,属于 OSOPER 的用户可直接以 SYSOPER 身份登录数据库。默认 OSDBA 组为 dba,OSOPER 组为 oinstall(如果创建了组 oper,则此处为 oper)。

图 1-38 显示了 Oracle 软件安装前的系统环境检查进度。

图 1-39 显示了软件安装前的系统环境检查结果,未满足安装要求的项目要逐一检查,对于包版本与要求不一致的可忽略,对于缺少的包需要单独安装,安装后再重新执行检查,直到满足要求为止。此处选择 Ignore All 复选框。

图 1-40 所示的页面显示了安装的概要信息,如果没有问题则单击 Finish 按钮进行软件安装。此处单击 Finish 按钮。如图 1-41 所示的页面显示了安装进度。

图 1-42 所示的页面显示了软件安装完成前需要执行的脚本,按要求以 root 用户登录,分别执行这两个脚本,执行成功后单击 OK 按钮,最后显示如图 1-43 所示的页面,安装成功。

图 1-36 产品清单目录

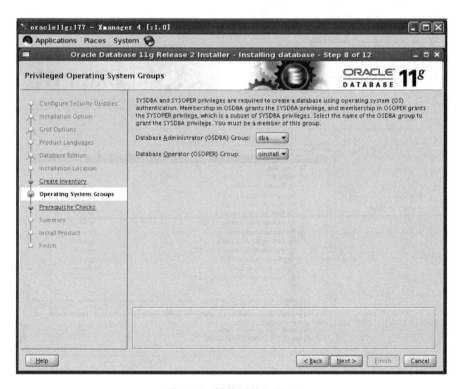

图 1-37 操作系统组设置

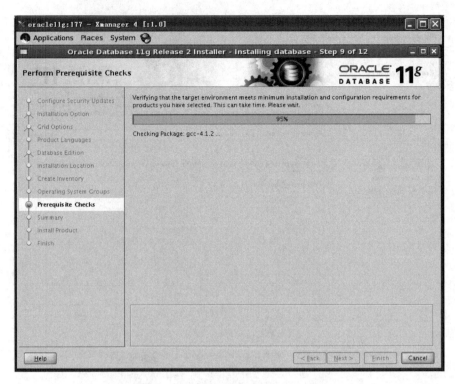

图 1-38　软件安装环境检查界面

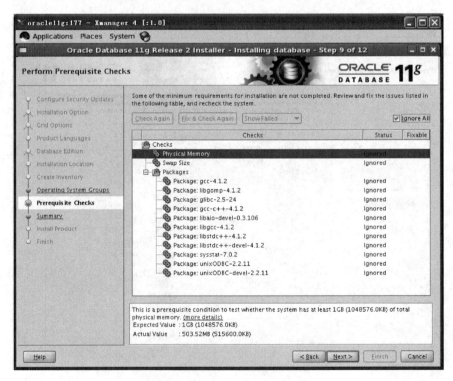

图 1-39　先决条件检查结果页面

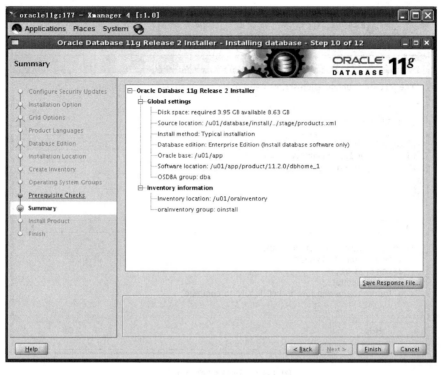

图 1-40　安装概要信息页面

图 1-41　安装进度

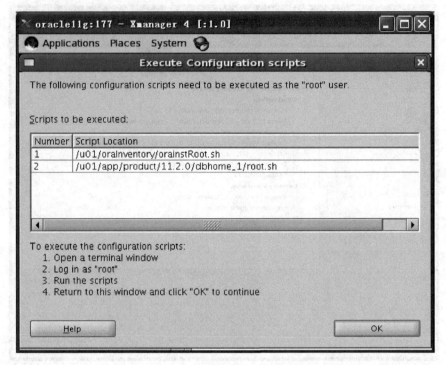

图 1-42　执行配置脚本提示

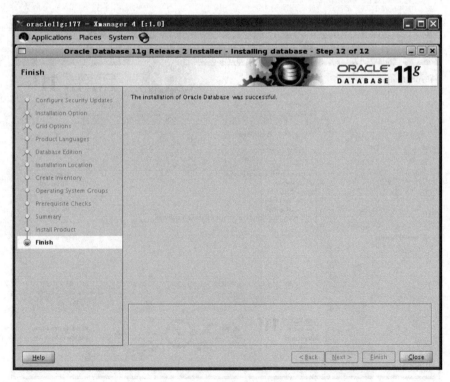

图 1-43　安装成功页面

1.3 配置数据库监听器

1.3.1 使用 NETCA 配置监听器

监听器(Listener)是 Oracle 数据库接收客户端连接请求的进程,如果数据库提供非本地的连接操作,则必须要配置监听器。配置监听器可通过执行如下命令完成:

```
$ netca
```

命令执行后,会有如图 1-44 所示的配置界面出现,选择 Listener configuration 单选按钮,单击 Next 按钮。

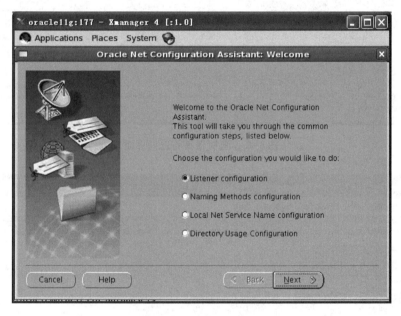

图 1-44 网络配置界面

在图 1-45 所示的显示监听器操作页面中,可对监听器进行添加(Add)、重新配置(Reconfiure)、删除(Delete)和重新命名(Rename)。此处选择 Add 单选按钮后,单击 Next 按钮。

图 1-46 所示的是显示监听器名称输入页面,默认为 LISTENER,建议不要修改,然后单击 Next 按钮。

图 1-47 所示的是显示监听器协议选择页面,默认为 TCP,建议不要修改,然后单击 Next 按钮。

图 1-48 所示的是显示监听器端口设置页面,Oracle 数据库监听器默认使用端口是 1521,建议不要修改,然后连续单击页面上的 Next 按钮,直到配置结束页面出现,如图 1-49 所示,单击 Finish 按钮结束配置。

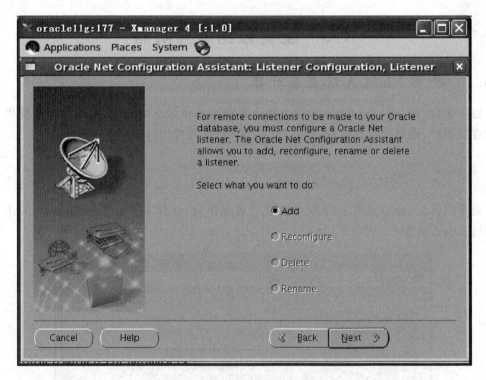

图 1-45　监听器操作选择页面

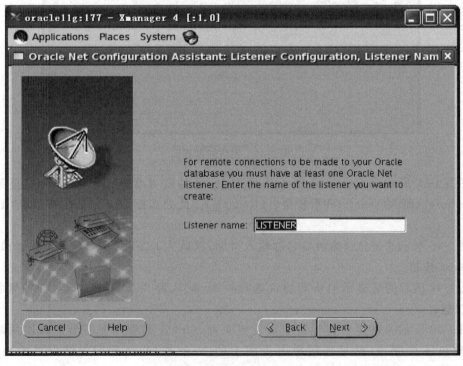

图 1-46　设置监听器名称

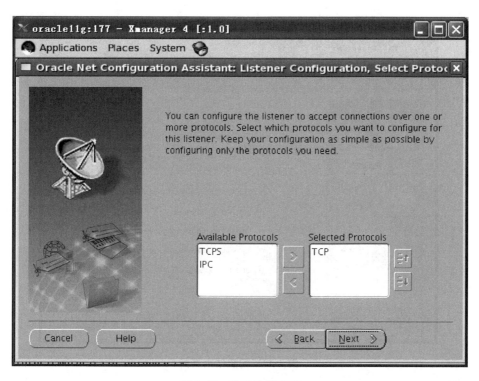

图 1-47　协议选择页面

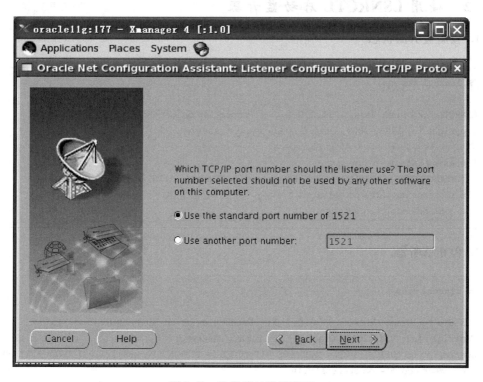

图 1-48　监听端口设置页面

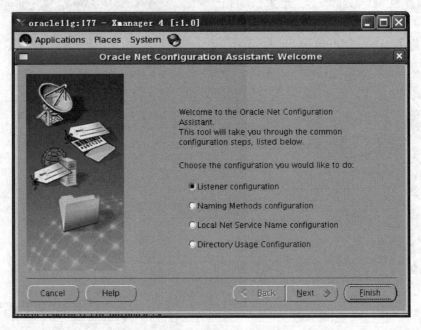

图 1-49 配置结束页面

1.3.2 使用 LSNRCTL 启动监听器

1. 查看监听器状态

```
$ lsnrctl status

LSNRCTL for Linux: Version 11.2.0.1.0 - Production on 21-JUN-2013 15:30:40
Copyright (c) 1991, 2009, Oracle.  All rights reserved.
...
Listening Endpoints Summary...
    (DESCRIPTION = (ADDRESS = (PROTOCOL = tcp)(HOST = oracle11g)(PORT = 1521)))
The listener supports no services
The command completed successfully
```

2. 停止监听器

```
$ lsnrctl stop

LSNRCTL for Linux: Version 11.2.0.1.0 - Production on 21-JUN-2013 15:30:45
Copyright (c) 1991, 2009, Oracle.  All rights reserved.
Connecting to (DESCRIPTION = (ADDRESS = (PROTOCOL = TCP)(HOST = oracle11g)(PORT = 1521)))
The command completed successfully
```

3. 启动监听器

```
$ lsnrctl start

LSNRCTL for Linux: Version 11.2.0.1.0 - Production on 21-JUN-2013 15:30:48
Copyright (c) 1991, 2009, Oracle.  All rights reserved.
Starting /u01/app/product/11.2.0/dbhome_1/bin/tnslsnr: please wait...

…

Listener Parameter File    /u01/app/product/11.2.0/dbhome_1/network/admin/ listener.ora
Listener Log File /u01/app/diag/tnslsnr/oracle11g/listener/alert/log.xml
Listening Endpoints Summary...
    (DESCRIPTION = (ADDRESS = (PROTOCOL = tcp)(HOST = oracle11g)(PORT = 1521)))
The listener supports no services
The command completed successfully
```

1.4 创建数据库

创建数据库有两种方法：①使用 SQL 脚本（CREATE DATABASE……）；②使用数据库配置助手 DBCA（Database Configuration Assistant）。此处使用 DBCA 来完成数据库创建：

```
$ dbca
```

命令执行后，出现如图 1-50 所示的欢迎页面，单击 Next 按钮。如图 1-51 所示页面显示数据库操作选项，可以创建数据库（Create a Database）、配置数据库选项（Configure Database Options）、删除数据库（Delete a Database）以及管理数据库模板（Manage Templates）。此处选择（Create a Database）单选按钮。

图 1-52 所示的是显示可选择的数据库模板，包括一般事务处理系统（General Purpose or Transaction Processing）（适用于一般的 OLTP 系统）、定制数据库（Custom Database）和数据仓库（Data Warehouse）（适用于一般的 OLAP 系统）。此处选择 Custom Database 单选按钮。

如图 1-53 所示页面显示数据库名称和实例名称的输入界面，这两个名称默认是相同的，但也可以不同。为区分数据库的这两个概念，此处数据库名为 mydb，实例名为 mysid。

图 1-54 所示的是显示 Enterprise Manager（EM）和自动维护任务设置页面。EM 是 Oracle 提供的基于 Web 页面的图形管理工具，此处暂时不进行 EM 配置，自动维护任务是 Oracle 数据库预定义的一些任务，可以定时统计表数据、自动进行 SQL 优化等，此处采取默认设置。

图 1-55 所示的是显示用户 sys 和 system 口令设置页面，sys 用户是 Oracle 数据库具有 SYSDBA 权限的超级用户，system 是具有 DBA 角色的数据库管理员用户，这两个用户的口令可单独设置，也可设置成相同的，此处为简便设置成相同口令。

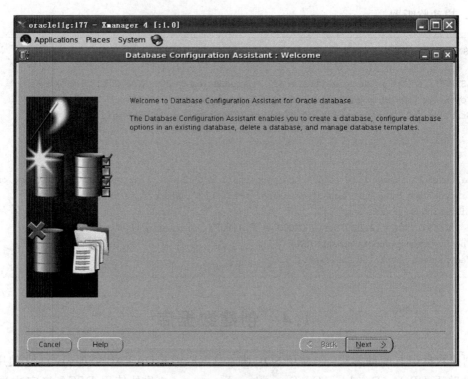

图 1-50　创建数据库欢迎页面

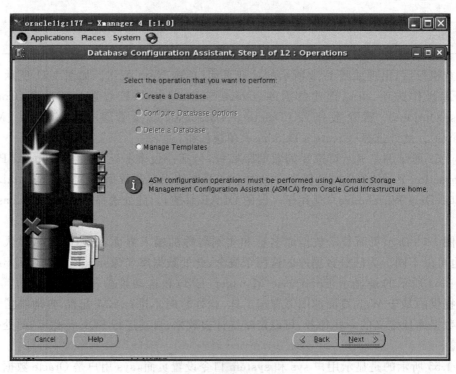

图 1-51　数据库操作选择

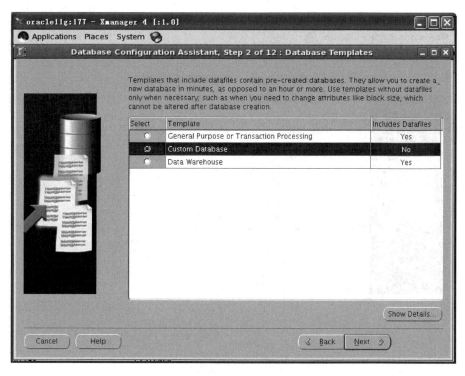

图 1-52 数据库模板选择页面

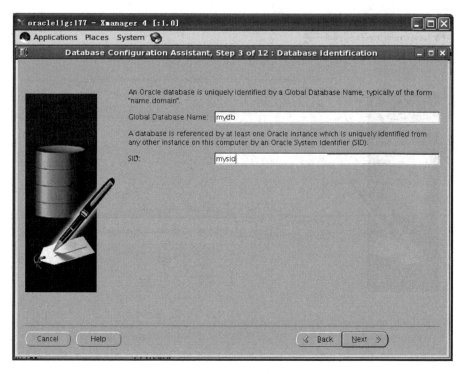

图 1-53 数据库和实例输入页面

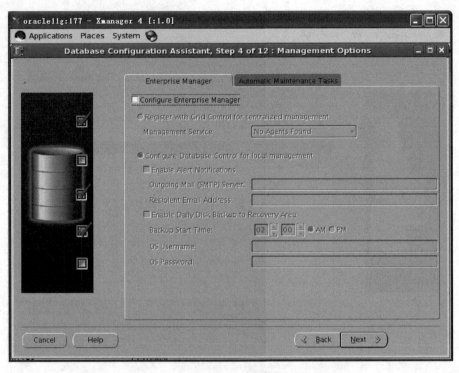

图 1-54 EM 和自动维护任务设置页面

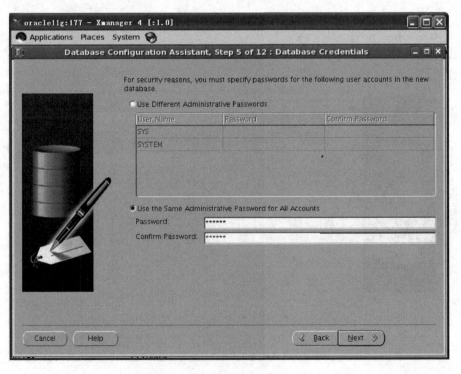

图 1-55 设置用户口令页面

图 1-56 所示的是显示数据库文件存储类型选择页面,其中 File System 指使用文件系统进行数据库文件管理,Automatic Storage Management(ASM)是从 Oracle 10g 开始提供的存储方式,由 Oracle 数据库特殊的 ASM 实例进行数据库文件存储的管理,此处选择 File System。

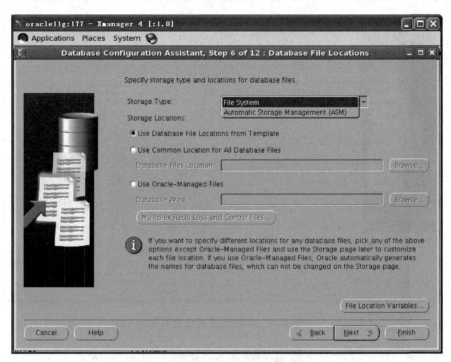

图 1-56　数据库文件存储类型选择页面

图 1-57 所示的是显示闪回恢复区设置页面,闪回恢复区可用来存储数据库备份文件、数据库闪回日志等内容,通过 Flash Recovery Area 设置闪回恢复区存储目录,通过 Flash Recovery Area Size 设置闪回恢复区使用空间的限制。

图 1-58 所示的是显示数据库组件选择页面,可根据需要进行组件选择,此处没有选择任何组件。另外单击 Standard Database Components 按钮,在如图 1-59 所示的页面中去掉所有选择项。

图 1-60 所示的是显示内存大小设置页面,选中 Use Automatic Memory Management 复选框,完全依赖数据库进行自动内存调整。

图 1-61 所示的是显示默认数据库块大小的设置,对于 OLTP 系统默认 8192B 完全可满足,如果是 OLAP 系统可适当增大块的大小,进程数指可启动的前台和后台进程的总数,根据客户端数量进行估计并设置。

图 1-62 所示的是显示数据库字符集选择页面,Database Character Set 设置数据库默认字符集,char 和 varchar2 类型的数据存储使用该字符集；National Character Set 设置国家字符集,nchar 和 nvarchar2 类型的数据存储使用该字符集。此处 Database Character Set 选择的是 AL32UTF8,National Character Set 选择的是 AL16UTF16。

如图 1-63 所示页面显示数据库服务器模式选择页面,Dedicated Server Mode 是专用服务器模式,Shared Server Mode 是共享服务器模式,此处选择 Dedicated Server Mode 单选按钮。

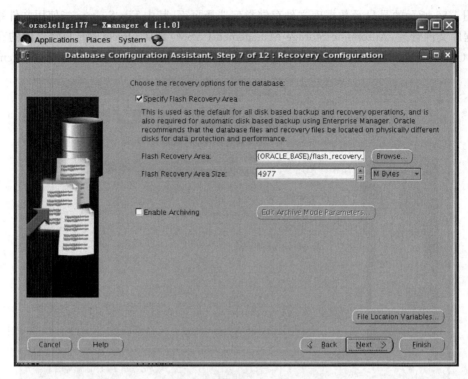

图 1-57　闪回恢复区设置页面

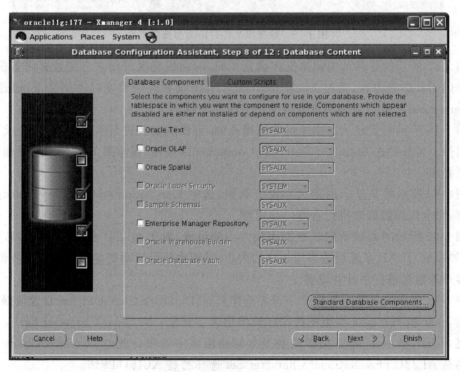

图 1-58　数据库组件选择页面

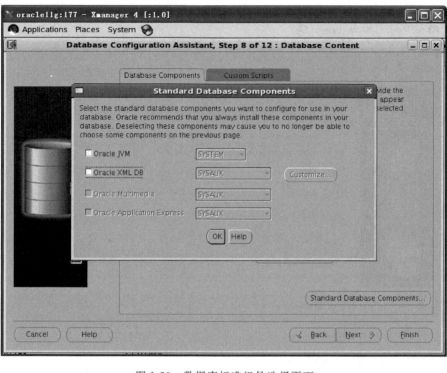

图 1-59　数据库标准组件选择页面

图 1-60　内存大小设置

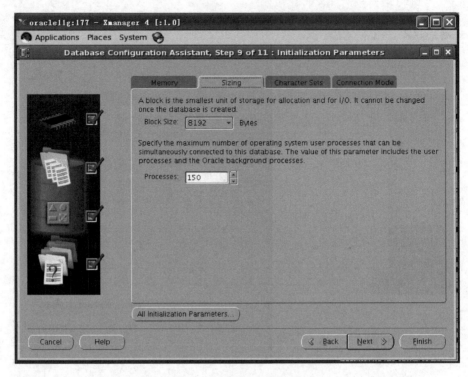

图 1-61　默认块和进程数设置

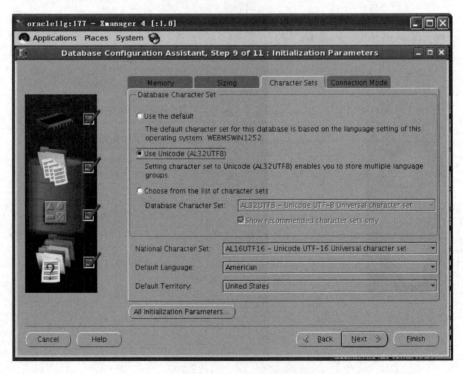

图 1-62　数据库字符集选择页面

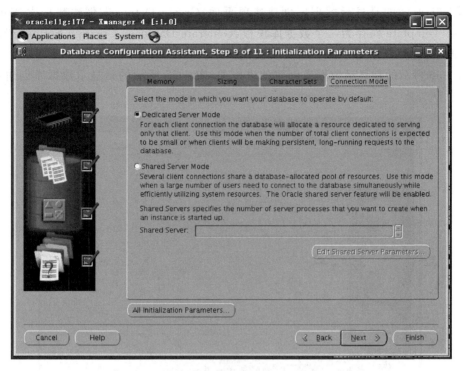

图 1-63　数据库连接模式选择页面

如图 1-64 所示页面显示数据库存储信息页面，可调整数据文件大小等，此处未作任何调整。

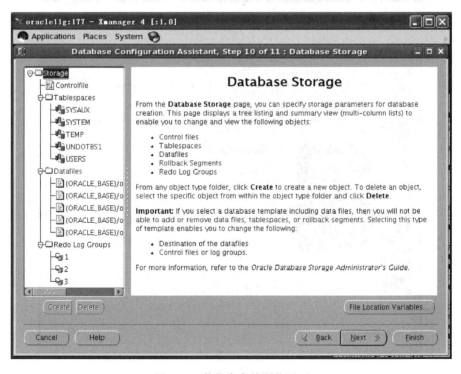

图 1-64　数据库存储浏览页面

图 1-65 所示的是显示数据库创建选项页面，Create Database 创建数据库，Save as a Database Template 保存为一个数据库模板，Generate Database Creation Scripts 产生创建数据库脚本，此处只选 Create Database 复选框。

图 1-66 所示的是显示数据库创建的进度条，图 1-67 所示的是显示数据库创建结束的提示信息。

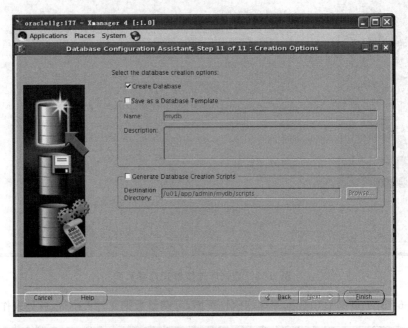

图 1-65　数据库创建选项选择页面

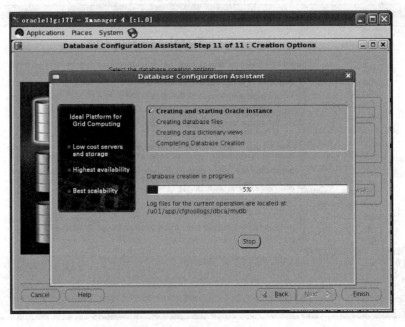

图 1-66　数据库创建进度

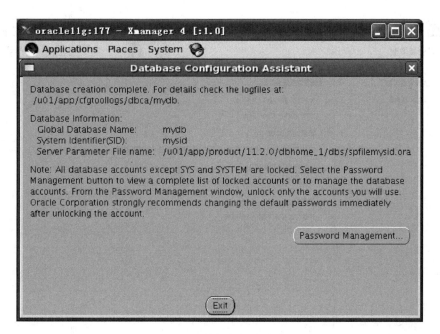

图 1-67　数据库创建结束

1.5　配置客户端

1.5.1　安装 Oracle 11g 客户端软件

Oracle 11g 数据库服务器已经安装在 Linux 系统,客户端选择 Windows XP 系统。客户端连接数据库,需要在客户端机器上安装 Oracle 11g 客户端软件。执行安装软件目录中的命令 setup:

```
C:\client> setup
```

如图 1-68 所示的是显示客户端安装类型,此处选择【管理员(1.02GB)】单选按钮,单击【下一步】按钮。图 1-69 所示的页面中显示了产品使用的语言,默认选择了【简体中文】和【英语】,单击【下一步】按钮。

图 1-70 所示的页面显示了客户端软件安装路径,【Oracle 基目录】指的是 ORACLE_BASE,【软件位置】指的是 ORACLE_HOME,此处采用默认值,单击【下一步】按钮。图 1-71 所示的页面显示了安装前环境检查结果,对未满足的检查项全部忽略,选中【全部忽略】复选框,单击【下一步】按钮。

图 1-72 所示的页面显示了客户端安装的基本信息设置,单击【下一步】按钮。图 1-73 所示的页面显示了安装进度条。安装结束后,显示如图 1-74 所示的页面,可单击【关闭】按钮。

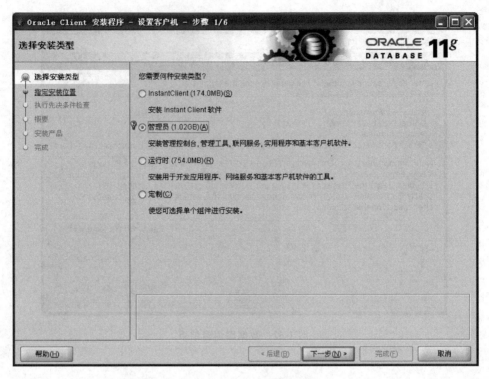

图 1-68　客户端安装类型

图 1-69　选择产品语言

图 1-70　选择安装目录

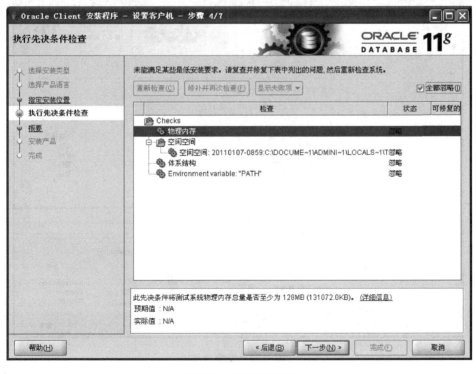

图 1-71　安装要求检查

图 1-72 安装信息浏览

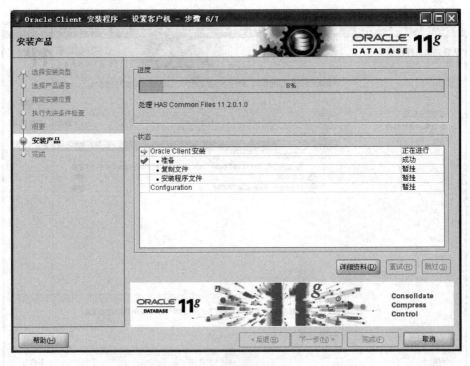

图 1-73 安装进度

图 1-74　安装完成

1.5.2　使用 NETCA 配置网络服务命名

客户端连接远程数据库需要配置服务命名,包括 IP 地址信息、监听端口信息以及数据库服务名信息。此处客户端环境为 Windows XP 系统,通过 NETCA 进行配置。

在"运行"对话框中输入 netca,如图 1-75 所示。

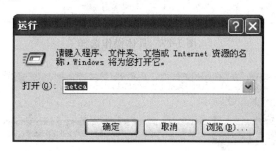

图 1-75　"运行"对话框

或者选择【开始】|【程序】| Oracle - OraClient11g_home1 |【配置和移植工具】| Net Configuration Assistant。

在如图 1-76 所示的页面中选择【本地网络服务名配置】单选按钮,单击【下一步】按钮。在如图 1-77 所示的页面中选择网络服务名的操作,此处选择【添加】单选按钮,单击【下一步】按钮。在如图 1-78 所示的页面中输入要连接的远程数据库的数据库名称 mydb,单击【下一步】按钮。

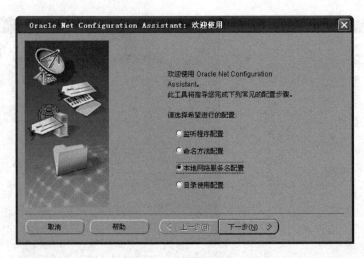

图 1-76　配置项选择

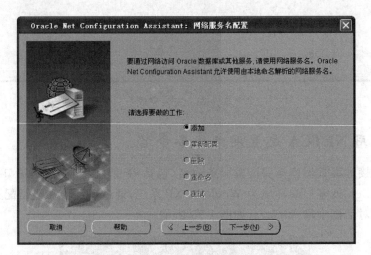

图 1-77　具体操作选择

图 1-78　输入数据库服务名

在如图 1-79 所示的页面中选择网络协议,此处选 TCP,单击【下一步】按钮。在如图 1-80 所示的页面中输入要连接的远程数据库的主机名或 IP 地址,此处输入 IP:192.168.1.2,如果输入主机名,需事先查看是否可以 ping 通,如果不通需要配置 C:\WINDOWS\ system32\ drivers\ etc\hosts 文件,将 IP 和主机名对应关系配置好,单击【下一步】按钮。在如图 1-81 所示页面中选择【是,进行测试】单选按钮,单击【下一步】按钮。

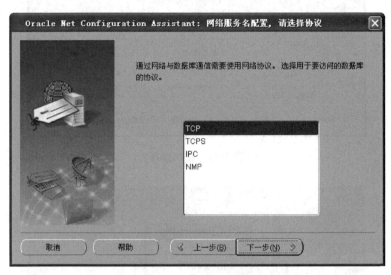

图 1-79 网络协议选择

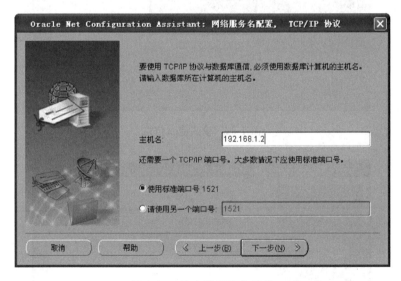

图 1-80 输入数据库主机名

在如图 1-82 所示的页面中单击【更改登录】按钮,在弹出的"更改登录"对话框中输入用户名 system,口令 oracle,单击【确定】按钮,接着显示如图 1-83 所示的测试成功页面,单击【下一步】按钮。在如图 1-84 所示的页面中输入要保存的网络服务名,可自行定义,此处输入 remote,接下来一直单击【下一步】按钮,直至出现如图 1-85 所示页面,单击【完成】按钮结束网络服务名的配置。

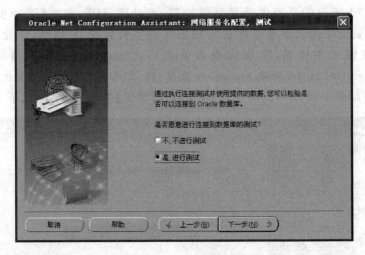

图 1-81 测试选择

图 1-82 输入登录信息

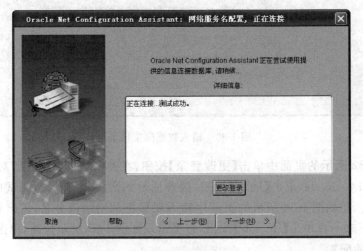

图 1-83 测试成功

图 1-84　输入网络服务名

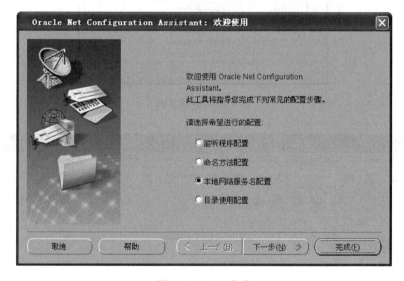

图 1-85　配置完成

1.5.3　连接数据库

连接 Oracle 数据库常用的工具有 SQL * Plus、PL/SQL Developer、Toad 以及 SQL Developer。其中 SQL * Plus 是 Oracle 数据库安装包自带的客户端工具，运行稳定，功能也很强；PL/SQL Developer 和 Toad 是业内使用较多的第三方软件公司开发的图形客户端工具；SQL Developer 是 Oracle 公司提供的图形客户端软件，在 Oracle 11g 中默认随软件一起安装，如果使用 Oracle 10g 或以前版本，可在 Oracle 公司官网上免费下载安装使用。

使用 SQL * Plus 连接数据库的命令为：

```
C:\> sqlplus system/oracle@remote

SQL * Plus: Release 11.2.0.1.0 Production on Tus Jun 25 09:43:29 2013
Copyright (c) 1982, 2010, Oracle.   All rights reserved.
Connected to:
Oracle Database 11g Enterprise Edition Release 11.2.0.1.0 - Production
With the Partitioning, OLAP, Data Mining and Real Application Testing options
SQL>
```

使用 PL/SQL Developer，图 1-86 所示的是显示其登录页面，Username 项输入用户名，Password 项输入用户口令，Database 项选择要连接远程数据库的网络服务名，Connect as 项选择连接数据库方式，普通用户选择 Normal。登录成功后，显示如图 1-87 所示的主页面。

图 1-86　登录页面

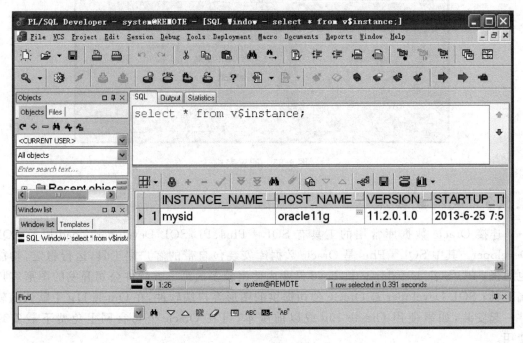

图 1-87　PL/SQL Developer 主页面

小　　结

　　学习 Oracle 数据库的第一步就是安装配置，用户的生产环境不同，Oracle 数据库的安装配置也会有很大差异。本章详细讲解了 Linux 环境下 Oracle 数据库的安装配置，包括 4 个方面内容：一是环境搭建，从 Oracle 数据库安装软件下载讲起，讲解虚拟机的创建、Red Hat Enterprise Linux 5 的安装配置及 Oracle 数据库软件的安装；二是数据库服务器端的监听器配置，包括配置工具 NETCA 使用和监听器控制工具 LSNRCTL 使用；三是详细介绍使用 DBCA 创建 Oracle 数据库；四是客户端配置，包括客户端 Oracle 软件安装、网络服务名配置及客户端软件的使用。

思　考　题

1. Linux 环境下安装 Oracle 数据库，需要建立哪些组？这些组的作用是什么？
2. 如何启动和停止 Oracle 数据库的监听器？
3. 从客户端连接 Oracle 数据库，需要提供哪些信息方能连接？

第二部分

Oracle数据库日常管理

第 2 章　数据库体系结构

了解 Oracle 数据库体系结构、掌握 Oracle 数据库的内部运行机制是更好地使用 Oracle 数据库的前提,下面将详细介绍 Oracle 数据库和实例的各个组成部分及其功能。

2.1　数据库体系结构简介

一个 Oracle 数据库服务器包含一个数据库和至少一个数据库实例,数据库和数据库实例密不可分,因而大多时候人们所说的 Oracle 数据库包含这两部分。在 RAC(Real Application Cluster)环境下一个数据库可有多个数据库实例来访问,但一个实例只能服务一个数据库。

1. 数据库

数据库(Database)是指存储数据的仓库,具体表现为存储数据的文件集合。

2. 数据库实例

数据库实例(Instance)负责数据库中数据操作,包含一组后台进程和一个大的共享内存区(System Global Area,SGA)。

图 2-1 显示了 Oracle 数据库的体系结构,包括用户进程(Client process)、服务器进程(Server process)、数据库实例(Instance)和数据库(Database)。其中,用户进程指连接数据库的客户端进程,可能是客户端工具或是程序;服务器进程指客户端连接数据库服务器后,

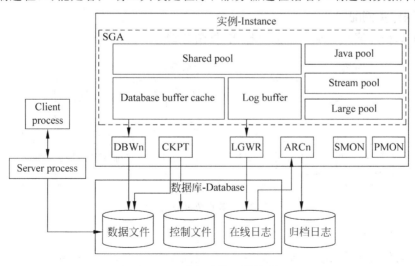

图 2-1　数据库体系结构图

在数据库服务器启动的进程,用于与客户端进程通信、处理客户端进程发送的 SQL 语句;数据库实例包括 SGA 和一组后台进程,用于完成数据库的管理;数据库包括数据文件、控制文件和日志文件,如果是归档模式,日志切换后由归档进程进行归档,生成归档日志。

2.2 数据库实例结构

2.2.1 数据库实例简介

1. 数据库实例介绍

在没有数据库服务器可用的情况下,是软件工程师通过编写程序来读写磁盘上的文件,有了数据库服务器,磁盘上的文件不再需要编写程序来直接访问,而是通过向数据库服务器发送 SQL(Structured Query Language)语句,数据库服务器接受 SQL 语句后解析,最终完成磁盘上文件的读写。

数据库实例的功能简单理解就是帮助读写磁盘文件。数据库实例由进程和内存组成,其中进程负责数据库文件管理和数据管理,而内存主要用来进行数据缓冲提高访问效率。当应用连接 Oracle 数据库时,连接的是数据库实例,实例启动前台的服务器进程来为应用服务。

2. 数据库实例结构

当 Oracle 数据库实例启动时,将分配一块大的内存区 SGA,并启动多个后台进程。SGA 的主要作用包括:

(1)维护多个数据库进程访问的内部数据结构。
(2)缓存从磁盘读出的数据块。
(3)缓存尚未写出到日志文件的日志数据。
(4)存储 SQL 的实行计划。

SGA 是被 Oracle 的进程所共享的,包括后台进程和服务于客户端的服务器进程,SGA 只能存在一台机器。Oracle 数据库实例结构可参考图 2-1。

3. 数据库实例配置

Oracle 数据库实例可运行在如下模式(见图 2-2)。

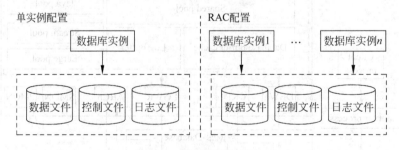

图 2-2 数据库实例模式

1) 单实例配置

数据库实例和数据库是一对一关系,一个数据库只有一个实例访问。

2）RAC 配置

数据库实例和数据库是多对一关系,一个数据库同时有多个实例访问,提高可用性,防止单个实例故障影响数据库使用,只要有一个实例还处于可用状态即不影响数据库的使用。

4. 检查点介绍

检查点对于一致数据库的 shutdown、实例恢复等都是非常重要的机制。检查点含义如下。

(1) 指示检查点位置,记录下一个实例恢复应用 redo 日志的起始 SCN 号。

(2) 将数据缓冲区中的数据写到磁盘。

检查点的意义:

(1) 减少实例恢复需要的时间。

(2) 确保有规律地将数据缓冲区中的脏数据块写到磁盘数据文件中。

(3) 确保一致 shutdown 数据库时所有提交的数据写到磁盘数据文件。

检查点分类:

1) 完全检查点

数据库写所有的数据缓冲区数据到磁盘,发生条件有:一致性 shutdown 数据库;手动执行命令 alter system checkpoint;在线日志切换;执行数据库热备份 alter database begin backup。

2) 部分检查点

将部分数据文件在数据缓冲区中的数据写到磁盘,发生条件有:执行表空间热备份 alter tablespace xxx begin backup;表空间只读 alter tablespace xxx read only;表空间离线 alter tablespace xxx offline;数据文件收缩。

3) 增量检查点

通过设置系统参数(例如 fast_start_mttr_target),数据库增量写出缓冲区中的脏数据。DBWn 进程每 3s 检查一次是否有数据需要写出,如有则写出数据并更新控制文件中检查点位置。

5. 实例恢复介绍

实例恢复是应用日志重新构造一致性数据库的过程,实例恢复发生在数据库不一致关闭后的重新启动,由 SMON 进行负责恢复,不需要人为参与。

数据库的数据变化都是通过日志写出,任何事务提交都会触发 LGWR 将 Log Buffer 中日志条目和事务 SCN 号写出到日志文件。同时 DBWn 也会在内存空间不足时将修改的数据块写出到数据文件。这时就可能造成数据文件中包含未提交的数据和不包含已提交的数据。

数据库在 shutdown abort 或异常中断后,可能导致如下情况发生。

(1) 事务提交写到了在线日志文件但没写到数据文件,这些事务变化必须应用到数据文件。

(2) 数据文件包含未提交事务的数据,这些数据必须回退。

实例恢复过程(如图 2-3 所示):

(1) 前滚(rolling forward)。实例恢复的第一阶段称为前滚,应用上次检查点后的所有日志。前滚后,数据文件将包含上次检查点后的所有提交数据和未提交的数据。

（2）回滚（rolling back）。实例恢复的第二阶段称为回滚，应用 undo 信息将未提交的数据全部回退。

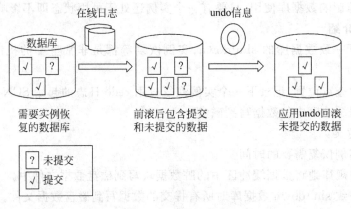

图 2-3 实例恢复过程

2.2.2 实例内存结构

1. 主要内存结构

Oracle 数据库的主要内存结构如下。

1）系统全局区

SGA 是一组共享的内存结构，包含数据库的控制信息和数据信息。SGA 被所有的服务器进程和后台进程所共享。

2）程序全局区

PGA（Program Global Area）是非共享的内存区，包含的控制信息和数据信息专属于一个 Oracle 数据库进程。

2. 内存管理方式

Oracle 数据库管理内存基于内存相关的初始化参数设置，管理方式包括以下三种。

1）自动内存管理（Automatic Memory Management，AMM）

通过设置系统参数 memory_target，指定数据库实例可用的内存大小，由 Oracle 数据库实例自动调整各个内存区的大小，根据需要重新分配 SGA 和 PGA 内存区的大小。

2）自动共享内存管理（Automatic Shared Memory Management，ASMM）

通过设置系统参数 sga_target 和 pga_aggregate_target 分别指定 SGA 和 PGA 的大小，Oracle 数据库实例可在 SGA 内部根据需要动态调整各个组件分区的大小，以及动态调整各个进程占用的 PGA 的大小。

3）手动内存管理（Manual Memory Management）

手动内存管理分别设置 SGA 组件分区和 PGA 组件分区的大小。

3. PGA 介绍

PGA 是每个进程的私有内存区域，所有进程的 PGA 组成数据库实例的 PGA，如图 2-4 所示。

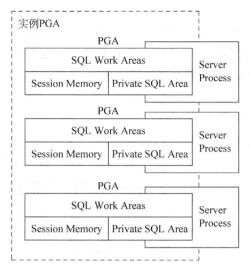

图 2-4 实例 PGA

每个进程的 PGA 分为不同的区域，如图 2-5 显示了专有服务器(dedicated server)模式下的 PGA 的组成部分。

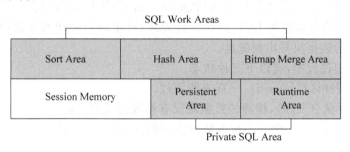

图 2-5 PGA 组成

1）私有 SQL 区

私有 SQL 区(Private SQL Area)持有解析的 SQL 语句信息和会话相关的信息。当服务器进程执行 SQL 或 PL/SQL，该区域用来存储绑定变量值、查询执行状态等信息。私有 SQL 区分为两个区域：运行时区域(run time area)，这个区域包含查询执行状态信息，比如全表扫描时目前为止获取的行数；持久区域(persistent area)，这个区域包含绑定变量值，供 SQL 执行时使用。

2）SQL 工作区

SQL 工作区(SQL Work Areas)用来进行数据在内存中的操作，比如数据排序、表数据连接等。SQL 工作区又细分了几个区域：排序区(Sort Area)用来执行数据的排序，只要 SQL 中包含 order by、distinct、group by 等语句就会引发排序；哈希区(Hash Area)用来执行连接查询(hash join)；位图合并区(Bitmap Merge Area)用来合并来自多个位图索引的扫描数据。这些内存区在手动内存管理时都有对应的系统参数可以设置，sort_area_size 用来设置单个进程内排序区大小，hash_area_size 用来设置单个进程内哈希区大小，bitmap_merge_area_size 用来设置单个进程内位图合并区大小。

3）会话内存区

会话内存区(Session Memory)也称为 UGA(User Global Area)，用来存储和会话相关的信息，比如登录时间、登录用户名、客户端所在机器等。这部分内存空间在专有服务器模式下从 PGA 中分配，如果是共享服务器模式则在 SGA 中分配。

4. SGA 介绍

SGA 分为多个内存区，如图 2-6 所示。

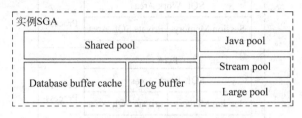

图 2-6　实例 SGA

1）数据缓冲池

数据缓冲池(Database Buffer Cache)用来存储从数据文件读取的数据块，主要作用是优化物理 IO，数据库修改数据时从数据文件中读取并存储在缓冲区中，数据修改在缓冲区中完成，如果事务提交，日志立即写出，但数据块延后写，等检查点发生时写出，可减少大量物理 IO，缓冲区中的数据使用 LRU 算法进行管理，对于频繁访问的数据块保留在缓冲区，对于不频繁访问的数据块写出到磁盘。

缓冲区在使用过程中，有以下三种状态。

(1) 未使用缓冲区(Unused)，一直未使用过。

(2) 干净缓冲区(Clean)，缓冲的数据块没有修改过或修改后已经写出到数据文件。

(3) 脏缓冲区(Dirty)，缓冲的数据块修改过但尚未写出到数据文件。

缓冲区分类：

(1) 默认池(Default Pool)，数据块存储默认使用的缓冲池。

(2) 保留池(Keep Pool)，可使频繁访问的数据块保留在内存中。

(3) 回收池(Recycle Pool)，对于不频繁访问的数据块放入该池中。

在 AMM 或 ASMM 管理方式下，数据库缓冲池大小由数据库自动调整，但也可以根据情况设置缓冲池的大小，这时数据库自动分配给该缓冲区的大小不低于参数设置。Oracle 数据库可同时支持多种块大小，创建数据库时默认是 8KB，针对多种不同大小的数据块需要分别设置相应块大小的缓冲区。数据缓冲池相关系统参数参见表 2-1。

表 2-1　内存相关参数

参　数	描　述
db_cache_size	默认池大小，默认值 0。如果默认数据库块大小为 8KB，则该缓冲池将存储 8KB 大小的数据库块，这时不能再设置 db_8k_cache_size。默认块大小缓冲池由系统自动设置，非默认块大小的缓冲池及保留池和回收池均需通过参数设置
db_2k_cache_size	2KB 数据块大小的缓冲池大小
db_4k_cache_size	4KB 数据块大小的缓冲池大小

续表

参数	描述
db_8k_cache_size	8KB 数据块大小的缓冲池大小
db_16k_cache_size	16KB 数据块大小的缓冲池大小
db_32k_cache_size	32KB 数据块大小的缓冲池大小
db_keep_cache_size	保留池大小
db_recycle_cache_size	回收池大小

【例】 查看并修改数据库缓冲区大小。

```
SQL > show parameter db_cache_size
NAME                            TYPE            VALUE
------------------------------  --------------  ------------------------------
db_cache_size                   big integer     0

SQL > alter system set db_cache_size = 40M;
System altered.

SQL > show parameter db_cache_size
NAME                            TYPE            VALUE
------------------------------  --------------  ------------------------------
db_cache_size                   big integer     40M
```

2）REDO 日志缓冲区

日志缓冲区（Redo Log Buffer）用于存储数据变化的日志记录，这些日志记录可用于重新构建丢失的数据。服务器进程将 PGA 中的日志记录复制到 SGA 中的日志缓冲区，后台进程 LGWR 将把日志缓冲区内容写到在线日志文件中，如图 2-7 所示。

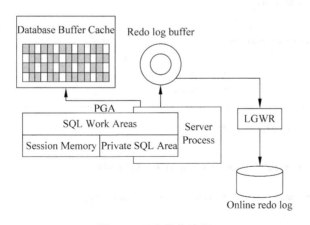

图 2-7　日志操作流程

【例】 查看日志缓冲区大小。

```
SQL > show parameter log_buffer
NAME                                 TYPE         VALUE
------------------------------------ ------------ ------------------------------
log_buffer                           integer      6135808
```

3) 共享池

共享池(Shared Pool)缓存了不同类型数据,比如解析过的 SQL、PL/SQL 代码、数据字典、执行结果等,包含的组件如图 2-8 所示。

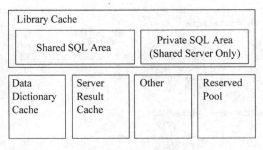

图 2-8 共享池组件

(1) 库缓存(Library Cache):存储可执行的 SQL 和 PL/SQL 代码。当一个新的 SQL 执行时,SQL 被解析后执行,这个解析过程称为硬解析(hard parse),解析过的 SQL 存储在库缓存中,当再次执行时将重用解析过的信息,这个过程称为软解析(soft parse)。

(2) 数据字典缓存(Data Dictionary Cache):存储数据字典信息,以行的形式存储数据,所以也称为 row cache。

(3) 结果缓存(Server Result Cache):存储 SQL 或 PL/SQL 的执行结果。

(4) 预留池(Reserved Pool):预留的一块内存,供有大内存分配时使用。

【例】 查看共享池大小。

```
SQL > show parameter shared_pool_size
NAME                                 TYPE         VALUE
------------------------------------ ------------ ------------------------------
shared_pool_size                     big integer  0
```

【例】 通过提示进行结果缓存。

```
SQL > select /* + result_cache */ count(*) from big;

SQL > select type, name, cache_id from v$result_cache_objects;
TYPE    NAME                                                       CACHE_ID
------  ---------------------------------------------------------  ----------------
Result  select /* + result_cache */ count(*) from big              7pfyhqpq8f3tmkvtv6

SQL > set autotrace traceonly exp
```

```
SQL> select /* + result_cache */ count(*) from big;
Execution Plan
----------------------------------------------------------

Plan hash value: 3110421800
----------------------------------------------------------
| Id  | Operation          | Name                  | Rows  | Cost ( %CPU) |
----------------------------------------------------------
|  0  | SELECT STATEMENT   |                       |     1 |   51 (0)     |
|  1  |  RESULT CACHE      | 7pfyhqpq8f3tmkvtv6    |       |              |
|  2  |   SORT AGGREGATE   |                       |     1 |              |
|  3  |    TABLE ACCESS FULL| BIG                  | 13135 |   51 (0)     |
----------------------------------------------------------

Result Cache Information (identified by operation id):
------------------------------------------------------
  1 - column-count = 1; dependencies = (U1.BIG); attributes = (single-row); name = "select
/* + result_cache */ count(*) from big"
```

4) 大池

大池(Large Pool)是可选的内存区,常用于大内存分配。

(1) 共享服务器模式下 UGA 存储。

(2) 并行查询。

(3) RMAN 备份恢复。

【例】 查看大池大小。

```
SQL> show parameter large_pool_size
NAME                                 TYPE         VALUE
------------------------------------ ------------ ------------------------------
large_pool_size                      big integer  0
```

5) Java 池

Java 池(Java Pool)是运行 Java 程序需要的内存空间,可选。

【例】 查看 Java 池大小。

```
SQL> show parameter java_pool_size
NAME                                 TYPE         VALUE
------------------------------------ ------------ ------------------------------
java_pool_size                       big integer  0
```

6）流池

流池（Streams Pool）应用于 Oracle 的流技术，用于存储队列消息。

【例】 查看流池大小。

```
SQL> show parameter streams_pool_size
NAME                                 TYPE              VALUE
------------------------------------ ----------------- ------------------------------
streams_pool_size                    big integer       0
```

2.2.3 Oracle 实例进程结构

1. 数据库实例相关进程

进程是操作系统执行一系列任务的机制，Oracle 数据库号称是多进程数据库服务器，包含以下几类进程。

（1）客户端进程（Client Processes）。指运行的应用程序或 Oracle 软件的客户端工具，比如 SQL*Plus。

（2）Oracle 数据库进程（Oracle Processes）。运行 Oracle 软件代码，包括两类：后台进程（Background Processes），随实例一起启动，完成实例恢复、写日志或写数据文件等进程；服务器进程（Server Processes），执行客户端请求。

2. 服务器进程

当客户端连接数据库时，数据库创建一个服务器进程来处理客户端的请求。服务器进程可执行如下任务。

（1）解析和运行 SQL 语句。

（2）执行 PL/SQL 代码。

（3）读数据文件，并将所读数据块放入数据缓冲区。

（4）返回结果。

服务器进程在专有服务器模式下（Dedicated Server）与客户端进程是一对一关系，如图 2-9 所示。在共享服务器模式下（Shared Server）与客户端进程是一对多关系，客户端进程与分发器进程（Dispatcher）通信，分发器进程收到请求放入请求队列，空闲的服务器进程从请求队列获取请求处理并将处理结果放入响应队列，分发器进程再从响应队列取回结果返给客户端进程，如图 2-10 所示。

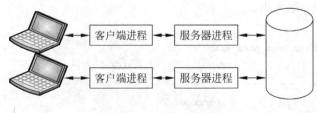

图 2-9 专有服务器模式

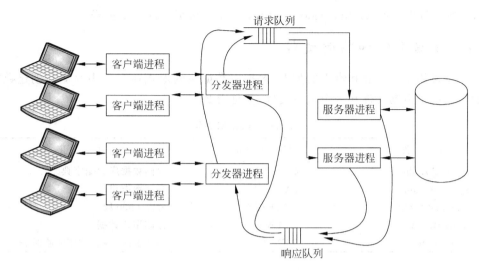

图 2-10 共享服务器模式

3. 后台进程

后台进程比较多,这里介绍几个最基本的后台进程。

1)进程监控器(PMON)

进程监控器监控后台进程,当进程终止时执行进程恢复;客户端进程异常结束时,负责释放客户端进程占用资源;向监听器进程注册数据库实例信息。

2)系统监控器(SMON)

系统监控器负责系统级的清理任务,包括实例恢复、清除临时空间,以及在字典管理表空间负责合并空闲区。

3)数据库写进程(DBWn)

在加载新数据出现数据缓冲区空间不足或检查点发生时,数据库写进程负责将数据缓冲区中的数据块写到磁盘文件。

4)日志写进程(LGWR)

日志写进程负责将日志缓冲区中日志记录写到在线日志文件中。

5)检查点进程(CKPT)

检查点进程负责更新控制文件和数据文件头中的检查点信息,并发信号通知 DBWn 写缓冲区中数据到磁盘文件。

6)归档进程(ARCn)

在日志切换发生时,归档进程复制在线日志文件内容到归档文件。

2.3 数据库物理结构

Oracle 数据库是一组文件的集合,包含以下三类文件。

(1)数据文件和临时文件(Data files and temp files)。数据文件包含数据库的数据存储结构,比如表和索引。临时文件用于临时表或数据磁盘排序等。

(2)控制文件(Control files)。记录数据库的结构信息,号称是数据库的"大脑"。

(3) 在线日志文件(Online redo log files)。记录数据库的改变记录。

2.3.1 数据文件和临时文件

数据文件用于存储表和索引的数据,数据以数据块的形式在文件中存储。创建数据库时一般包含四个数据文件和一个临时文件,参见表2-2。

表 2-2 数据库文件

文件	所属表空间	作用
system01.dbf	SYSTEM	存储数据库数据字典
sysaux01.dbf	SYSAUX	存储数据组件
undotbs01.dbf	UNDOTBS	分配 undo 段
users01.dbf	USERS	存储用户数据
temp01.dbf	TEMP	存储临时数据,排序或临时表

【例】 查看数据库数据文件和临时文件。

```
SQL > select name, bytes from v $ datafile;
NAME                                           BYTES
--------------------------------------------   ----------
/u01/app/oradata/mydb/system01.dbf             734003200
/u01/app/oradata/mydb/sysaux01.dbf             629145600
/u01/app/oradata/mydb/undotbs01.dbf            209715200
/u01/app/oradata/mydb/users01.dbf              5242880


SQL > select name, bytes from v $ tempfile;
NAME                                           BYTES
--------------------------------------------   ----------
/u01/app/oradata/mydb/temp01.dbf               20971520
```

2.3.2 控制文件

控制文件号称是数据库的"大脑",记录数据库状态和结构信息。数据库实例启动时读取参数文件,从参数(control_files)中得知控制文件位置,再读取控制文件获取数据库中的数据文件和日志文件位置,最后打开所有文件。控制文件中记录很多信息,主要包括:

(1) 数据库名称和数据库唯一标识 DBID。
(2) 数据库创建时间。
(3) 数据文件、在线日志文件和归档日志的信息。
(4) 表空间信息。
(5) RMAN 备份信息。

【例】 查看控制文件(多个控制文件设置是防止单个损坏)。

```
SQL > show parameter control_files;
NAME                    TYPE        VALUE
----------------------- ----------- ----------------------------------------
control_files string                /u01/app/oradata/mydb/control01.ctl,
                                    /u01/app/flash_recovery_area/mydb/control02.ctl

SQL > select name from v $ controlfile;
NAME
--------------------------------------------------------------------------
/u01/app/oradata/mydb/control01.ctl
/u01/app/flash_recovery_area/mydb/control02.ctl
```

2.3.3 在线日志文件

在线日志文件用来记录数据库的变化,一旦数据文件损坏,可使用日志文件内容进行数据恢复。

【例】 查看日志文件。

```
SQL > select member from v $ logfile;
MEMBER
--------------------------------------------------------------------------
/u01/app/oradata/mydb/redo01.log
/u01/app/oradata/mydb/redo02.log
/u01/app/oradata/mydb/redo03.log
```

2.4 参数文件管理

2.4.1 参数文件介绍和管理

数据库实例启动时需要读初始化参数文件,参数文件可以有两种形式:服务器参数文件(server parameter file)和文本初始化参数文件(text initialization parameter file)。首选使用服务器参数文件。

1. 服务器参数文件

服务器参数文件由 Oracle 数据库进行管理。服务器参数文件有如下 4 个特征。

(1) 一个数据库只有一个服务器参数文件,且必须驻留在数据库所在机器上。

(2) 服务器参数文件由数据库完成读写。

(3) 服务器参数文件是二进制文件,不能使用文本编辑器进行修改。

(4) 通过命令修改参数时可持久保留在服务器参数文件中。

服务器参数文件命名规则为:spfile + 实例名.ora,假设实例名为 ORCL,则对应的服务器参数文件名为 spfileORCL.ora,默认存储路径为 $ORACLE_HOME/dbs/(Windows 系统存储路径为%ORACLE_HOME%\database)。

2. 文本初始化参数文件

文本参数文件是以前版本使用的方式,目前 11g 版本默认使用服务器参数文件。文本参数文件有如下 4 个特征。

(1) 启停数据库时,文本参数文件必须驻留在发出命令的客户端。

(2) 是文本类型,不是二进制,可随意修改。

(3) Oracle 数据库只能读不能修改文本参数文件,修改参数必须使用文本编辑器人为修改。

(4) 通过命令修改参数只对当前实例起作用,手动修改文件后必须重新启动方可生效。

文本参数文件命名规则为:init + 实例名.ora。假设实例名为 ORCL,则对应的文本参数文件名为 initORCL.ora,默认存储路径为 $ORACLE_HOME/dbs/(Windows 系统存储路径为 %ORACLE_HOME%\database)。

3. 两种参数文件转换命令

服务器参数文件可转换成文本参数文件,命令如下:

```
SQL> create pfile from spfile;
```

文本参数文件可转换成服务器参数文件,命令如下:

```
SQL> create spfile from pfile;
```

4. 实例启动搜索参数文件顺序

数据库实例启动时搜索参数文件顺序如下(假设实例名为 mysid)。

(1) spfilemysid.ora。

(2) spfile.ora。

(3) initmysid.ora。

5. 服务器参数文件参数修改语法

语法如下:

```
alter system set pname = pvalue scope = { both | memory | spfile }
```

参数说明:

pname:参数名称,具体参数名参考 Oracle 的技术文档。

pvalue:参数值,对于固定选项的参数值,可参考 v$parameter_valid_values。

scope:指定参数修改的范围,both 指同时修改内存和参数文件,memory 指改内存但不修改参数文件,实例重新启动该修改失效,spfile 指只修改参数文件,重新启动后生效。

【例】 修改参数示例(示例中修改参数第一次失败,这个参数不是动态参数,只能修改参数文件,重新启动后生效)。

```
SQL > show parameter remote_login_passwordfile;
NAME                                   TYPE        VALUE
------------------------------------ ---------- ------------------
remote_login_passwordfile              string      EXCLUSIVE

SQL > select value from v $ parameter_valid_values
where name = 'remote_login_passwordfile';
VALUE
--------------------------------------------------------------
SHARED
EXCLUSIVE
NONE

SQL > alter system set remote_login_passwordfile = none;
alter system set remote_login_passwordfile = none
            *
ERROR at line 1:
ORA - 02095: specified initialization parameter cannot be modified

SQL > alter system set remote_login_passwordfile = none scope = spfile;
System altered.
```

2.4.2 基本参数介绍

【例】 转换服务器参数文件为文本参数文件,查看参数设置。

```
$ sqlplus / as sysdba
SQL > create pfile from spfile;
SQL > quit
$ cd $ ORACLE_HOME/dbs
$ more initmysid.ora
* .audit_file_dest = '/u01/app/admin/mydb/adump'
* .compatible = '11.2.0.1.0'
* .control_files = '/u01/app/oradata/mydb/control01.ctl'
* .db_block_size = 8192
* .db_domain = ''
* .db_name = 'mydb'
* .db_file_multiblock_read_count = 16
* .db_recovery_file_dest_size = 2147483648
* .db_recovery_file_dest = '/u01/app/flash_recovery_area'
* .diagnostic_dest = '/u01/app/diag'
* .memory_target = 314572800
* .open_cursors = 300
* .processes = 150
* .remote_login_passwordfile = 'EXCLUSIVE'
* .undo_management = 'AUTO'
* .undo_retention = 900
* .undo_tablespace = 'UNDOTBS1'
```

具体参数介绍见表 2-3。

表 2-3 参数说明

参 数 名	描 述
audit_file_dest	审计信息存储路径
compatible	数据库兼容版本号
control_files	数据库控制文件路径
db_domain	数据库域名
db_name	数据库名
db_file_multiblock_read_count	数据库多块读的块数,常用于全表扫描
db_recovery_file_dest_size	闪回恢复区大小
db_recovery_file_dest	闪回恢复区目录
diagnostic_dest	数据库诊断信息路径
memory_target	数据库实例使用内存总大小,是 SGA 和 PGA 之和
open_cursors	单个会话最大打开的游标数
processes	数据库最大的进程数
remote_login_passwordfile	远程登录口令文件使用设置
undo_management	Undo 管理方式,默认是自动
undo_retention	事务结束后 undo 信息保留时间
undo_tablespace	undo 表空间

2.5 口令文件管理

当用户以 SYSDBA 或 SYSOPER 权限访问数据库时,可用口令文件来验证身份,口令文件不能做普通用户身份验证。同时与系统参数 remote_login_passwordfile 配合使用,如果该参数设置为 none,则远程用户以 SYSDBA 或 SYSOPER 身份登录数据库时不能启停数据库。

创建口令文件命令语法如下:

```
$ orapwd file = $ ORACLE_HOME/dbs/文件名 password = 口令 entries = n force = { y | n }
```

命令参数说明:
1. file
file 指定口令文件名,口令文件默认存储在 $ORACLE_HOME/dbs 下,命名规则为 orapwd+实例名,该文件没有扩展名,假设实例名为 ORCL,则口令文件为 orapwORCL。如果为 Windows 环境,存储路径为%ORACLE_HOME%\database,文件命名规则为 pwd+实例名.ora,假设实例名为 ORCL,则口令文件为 pwdORCL.ora。
2. password
password 指定具有 SYSDBA 或 SYSOPER 权限的用户口令,当重建口令文件时,只有 SYS 用户具备 SYSDBA 和 SYSOPER 权限。
3. entries
entries 控制口令文件最多存储几个用户信息。

4. force

如果口令文件存在,可通过设置 force=y 强制创建。

【例】 远程启停数据库(注意参数 remote_login_passwordfile 变化)。

```
步骤1: 在 Linux 系统中创建口令文件
$ orapwd file = $ ORACLE_HOME/dbs/orapwmysid password = oracle force = y

步骤2: 在 Linux 系统中查看参数 remote_login_passwordfile 设置
$ sqlplus / as sysdba
SQL > show parameter remote_login_passwordfile
NAME                                 TYPE         VALUE
------------------------------------ ------------ ------------------------------
remote_login_passwordfile            string       NONE

步骤3: 在 Windows 系统中远程登录
C:\> sqlplus sys/oracle@remote as sysdba
SQL * Plus: Release 11.2.0.1.0 Production on Fri Jun 28 12:37:22 2013
Copyright (c) 1982, 2010, Oracle.  All rights reserved.
ERROR:
ORA - 01017: invalid username/password; logon denied

步骤4: 在 Linux 系统中修改参数 remote_login_passwordfile
$ sqlplus / as sysdba
SQL > alter system set remote_login_passwordfile = 'exclusive' scope = spfile;
System altered.
SQL > shutdown immediate
Database closed.
Database dismounted.
ORACLE instance shut down.

SQL > startup
ORACLE instance started.
Total System Global Area      313860096 bytes
Fixed Size                      1344652 bytes
Variable Size                 247466868 bytes
Database Buffers               58720256 bytes
Redo Buffers                    6328320 bytes
Database mounted.
Database opened.
SQL > show parameter remote_login_passwordfile
NAME                                 TYPE         VALUE
------------------------------------ ------------ ------------------------------
remote_login_passwordfile            string       EXCLUSIVE

步骤5: 在 Windows 系统中远程登录并完成停止数据库、启动数据库
C:\> sqlplus sys/oracle@remote as sysdba
SQL > shutdown immediate
Database closed.
Database dismounted.
```

```
ORACLE instance shut down.

SQL > startup
ORACLE instance started.
Total System Global Area    313860096 bytes
Fixed Size                    1344652 bytes
Variable Size               247466868 bytes
Database Buffers             58720256 bytes
Redo Buffers                  6328320 bytes
Database mounted.
Database opened.
```

2.6 数据库启动和停止

2.6.1 数据库启动

如图 2-11 所示，Oracle 数据库启动需要经历以下三个过程。

图 2-11 数据库启动过程

（1）启动实例

该过程只启动实例，不关联数据库。

（2）mount 数据库

实例已经启动，关联一个数据库，并打开该数据库的控制文件，但此时用户不能访问。

（3）open 数据库

实例已经启动，并根据控制文件中记录的数据库的结构信息，打开数据库的数据文件和日志文件，此时数据库可供用户访问。

1．startup 命令介绍

以 SYSDBA 或 SYSOPER 身份登录，输入 startup 命令即可启动数据库。实例启动时会根据环境变量 ORACLE_SID 设置的实例名称（假设 ORACLE_SID 为 xxx），在 $ORACLE_HOME/dbs 下查找 spfilexxx.ora 或是 initxxx.ora，然后启动。

```
$ sqlplus / as sysdba
SQL > startup
```

startup 命令常用参数说明如下。

nomount：启动数据库到 nomount 状态，这个过程读取参数文件，创建实例，分配内存，启动后台进程，然后可通过 alter database mount 命令进入 mount 状态。

mount：启动数据库到 mount 状态，这个过程读取控制文件，然后通过 alter database open 命令进入 open 状态。

open：启动数据库到 open 状态，默认的启动状态，这个过程对数据文件、日志文件进行检查点及完整性检查，如无问题打开文件，如有问题需要进行恢复。

pfile：指定启动参数文件。如未指定，将根据环境变量 ORACLE_SID 指定的数据库实例名在 $ORACLE_HOME/dbs 下进行搜索，搜索顺序：spfile 实例名.ora→spfile.ora→init 实例名.ora。

restrict：以限制方式打开，只有拥有 create session、restrict session 权限的用户才能连接数据库。可用于执行数据库数据导入导出、数据装载、暂停普通用户连接、数据库移植或升级。

force：不先用 shutdown 命令关闭数据库，而直接强行关闭并重启动。强制启动数据库，首先执行 shutdown abort，然后执行 startup，并进行实例恢复。也可使用 startup nomount force 或 startup mount force。

2. 启动实例过程

（1）在指定目录下（Linux 为 $ORACLE_HOME/dbs；Windows 为 %ORACLE_HOME%\database）搜索参数文件，先找二进制的参数文件 spfilemysid.ora（此处假设实例名为 mysid），如果未找到，找 spfile.ora，最后找文本类型的参数文件 initmysid.ora，如果仍未找到将报错。

（2）读取参数文件获取参数值。

（3）根据参数 memory_target 或 sga_target 的设置分配 SGA。

（4）启动后台进程。

（5）打开诊断路径下的 alert 日志文件和 trace 文件记录所有的参数设置。

【例】 找不到参数文件报错示例。

```
SQL> startup nomount
ORA-01078: failure in processing system parameters
LRM-00109: could not open parameter file '/u01/app/product/11.2.0/dbhome_1/dbs/initmysid.ora'
```

【例】 正常启动示例。

```
SQL> startup nomount
ORACLE instance started.

Total System Global Area    313860096 bytes
Fixed Size                    1336232 bytes
Variable Size               201329752 bytes
Database Buffers            104857600 bytes
Redo Buffers                  6336512 bytes

SQL> select status from v$instance;
STATUS
------------
STARTED
```

启动实例成功后,实例状态为 started,这个状态下可进行数据库创建、控制文件创建。

3. mount 数据库过程

mount 数据库时实例将关联到数据库,首先从参数文件中配置的参数 control_files 获取到数据库的控制文件,然后打开控制文件并读取控制文件中记录的数据文件和日志文件名。如果 mount 过程中发现控制文件丢失或多个控制文件不一致将报错。

如果数据库以 nomount 方式启动,mount 数据库时需执行 alter database mount 命令,如果数据库是 shutdown 状态,则直接执行 startup mount 命令。

【例】 找不到控制文件报错示例。

```
SQL> alter database mount;
alter database mount
*
ERROR at line 1:
ORA-00205: error in identifying control file, check alert log for more info
```

根据错误提示查看 alert 日志:

```
$ more /u01/app/diag/rdbms/mydb/mysid/trace/alert_mysid.log
...
alter database mount
Tue Jun 25 22:02:30 2013
ORA-00210: cannot open the specified control file
ORA-00202: control file: '/u01/app/flash_recovery_area/mydb/control02.ctl'
ORA-27037: unable to obtain file status
Linux Error: 2: No such file or directory
Additional information: 3
ORA-00210: cannot open the specified control file
ORA-00202: control file: '/u01/app/oradata/mydb/control01.ctl'
ORA-27037: unable to obtain file status
Linux Error: 2: No such file or directory
⋮
```

【例】 控制文件不一致报错示例。

```
SQL> alter database mount;
alter database mount
*
ERROR at line 1:
ORA-00214: control file '/u01/app/oradata/mydb/control01.ctl' version 442
inconsistent with file '/u01/app/flash_recovery_area/mydb/control02.ctl'
version 429
```

【例】 正常 mount 示例。

```
SQL > alter database mount;
Database altered.

SQL > select status from v $ instance;
STATUS
------------
MOUNTED
```

或直接启动到 mount 状态：

```
SQL > startup mount
ORACLE instance started.
Total System Global Area    313860096 bytes
Fixed Size                    1336232 bytes
Variable Size               201329752 bytes
Database Buffers            104857600 bytes
Redo Buffers                  6336512 bytes
Database mounted.
```

mount 数据库成功后，实例状态为 mounted，这个状态下可进行数据库维护，比如数据库恢复、数据库归档模式设置、数据库闪回特性设置等。

4. open 数据库过程

打开 mounted 状态数据库使数据库对所有用户可访问，需执行以下几步。

(1) 打开除 undo 表空间外所有表空间在线(online)状态的数据文件，如果表空间是离线(offline)状态，则该表空间和相应的数据文件仍旧保持离线状态，不予打开。

(2) 打开 undo 表空间对应的数据文件，如果存在多个 undo 表空间，则根据参数 undo_tablespace 打开指定的表空间，否则选择第一个可用的 undo 表空间。

(3) 打开所有的在线日志文件。

【例】 数据文件丢失报错示例。

```
SQL > alter database open;
alter database open
*
ERROR at line 1:
ORA - 01157: cannot identify/lock data file 4 - see DBWR trace file
ORA - 01110: data file 4: '/u01/app/oradata/mydb/users01.dbf'
```

【例】 数据文件旧报错示例。

```
SQL > alter database open;
alter database open
*
ERROR at line 1:
ORA - 01113: file 4 needs media recovery
ORA - 01110: data file 4: '/u01/app/oradata/mydb/users01.dbf'
```

Oracle Database 11g 管理与开发

【例】 正常 open 示例。

```
SQL> alter database open;
Database altered.

SQL> select status from v$instance;
STATUS
------------
OPEN
```

或直接启动到 open 状态：

```
SQL> startup
ORACLE instance started.

Total System Global Area   313860096 bytes
Fixed Size                    1336232 bytes
Variable Size               201329752 bytes
Database Buffers            104857600 bytes
Redo Buffers                  6336512 bytes
Database mounted.
Database opened.
```

2.6.2 数据库停止

如图 2-12 所示，Oracle 数据库停止需要经历以下三个过程。

图 2-12 数据库停止过程

（1）close 数据库

关闭数据文件和在线日志文件，数据库处于 mounted 状态。

（2）dismount 数据库

关闭数据库控制文件，断开与数据库关联关系，实例处于 started 状态。

（3）shutdown 实例

停止进程，释放分配内存。

1. shutdown 命令介绍

以 SYSDBA 或 SYSOPER 身份登录，输入 shutdown immediate 命令即可停止数据库。

```
$ sqlplus / as sysdba
SQL> shutdown immediate
```

shutdown 命令常用参数说明如下。

normal：默认方式。不允许新连接，等待所有当前连接退出。

immediate：不允许新连接，强制断开所有当前连接，回滚未提交事务。

transactional：不允许新连接、禁止新事务进行，允许当前事务完成，然后强制断开连接。

abort：立即终止所有连接、所有事务。这种方式在数据库重新启动时需要进行实例恢复。

shutdonw 命令选项见表 2-4。

表 2-4 shutdonw 命令选项

数据库行为	abort	immediate	transactional	normal
允许新连接	否	否	否	否
等待用户退出	否	否	否	是
等待事务完成	否	否	是	是
执行检查点并关闭文件	否	是	是	是

2. close 数据库过程

（1）数据库正常关闭（shutdown normal，shutdown transactional，shutdown immediate）。数据库将 SGA 中内容写出到数据文件和日志文件，接下来，关闭在线（online）数据文件和日志文件。这个状态下数据库对于普通用户是不可访问的。

（2）数据库异常关闭（shutdown abort 或异常宕机）。数据库实例突然停止，SGA 中的内容未写出，当重新启动数据库时需要实例恢复。

【例】 正常 close 数据库示例。

```
SQL> alter database close;
Database altered.

SQL> select status from v$instance;
STATUS
------------
MOUNTED
```

3. unmount 数据库过程

关闭数据库的控制文件，数据库实例保持 started 状态。

【例】 正常 unmount 数据库示例。

```
SQL > alter database dismount;
Database altered.

SQL > select status from v $ instance;
STATUS
------------
STARTED
```

4．shutdown 实例过程

停止实例，释放内存并停止所有后台进程。

【例】 正常 shutdown 实例示例。

```
SQL > shutdown ;
ORA - 01507: database not mounted
ORACLE instance shut down.
```

【例】 数据库 open 状态下正常 shutdown 数据库示例。

```
SQL > shutdown immediate
Database closed.
Database dismounted.
ORACLE instance shut down.
```

小　结

想要学好 Oracle 数据库，了解体系结构很重要。本章重点讲解了 Oracle 数据库系统的两个重要组成部分：一是实例，讲解了实例的主要进程和内存结构，并具体介绍了每个内存结构的作用；二是数据库，讲解了组成 Oracle 数据库的三类文件，即数据文件、控制文件和日志文件。

除了实例和数据库内容外，还讲解了参数文件、口令文件以及数据库的启停命令。参数文件部分，讲解了参数文件的类型、参数修改及常用参数的作用；口令文件部分，讲解了口令文件作用及创建方法；数据库启停部分，除讲解了命令使用，还讲解了 Oracle 数据库的启停过程。

思　考　题

1．Oracle 数据库系统中 SGA 和 PGA 分别是什么？
2．SGA 中主要包含哪些部分？描述每部分的作用。
3．Oracle 实例涉及哪些类进程？主要后台进程有哪些？
4．Oracle 数据库由哪些类文件组成？每种文件有什么作用？

5. 服务器参数文件和文本参数文件有何区别？
6. 将参数 MEMORY_TARGET 修改成 800MB，写出修改语句。
7. 创建一个口令文件，实例名称为 tsid，口令设成 oracle123，写出创建语句。
8. Oracle 数据库启动经过哪几个阶段？介绍每个阶段。
9. 关闭数据库有几种方式？简述其区别。

第 3 章 网 络 配 置

客户端应用程序和 Oracle 数据库之间通过网络进行连接，基于 Oracle 数据库的数据传输协议进行通信。本章重点讲解服务器端监听器和客户端命名解析文件的配置。

3.1 网络服务组件

数据在网络上传输时都需要遵守一定的规则，而这些规则常被称作协议。Oracle 数据库的客户端在与数据库服务器之间传递数据时需要遵守 Oracle 数据库的数据传输协议，Oracle 数据库软件的网络服务组件部分包含协议实现，这部分软件称为 Oracle Net。Oracle 数据库的网络服务部分除了 Oracle Net 外，还包括负责监听客户端连接的组件，这个组件称作 Oracle Net Listener，也就是人们所熟悉的监听器。

3.1.1 Oracle Net

Oracle Net 驻留在任何一台安装 Oracle 数据库服务器软件或客户端软件的机器上，负责建立和维护数据库客户端和服务器之间的连接。当一个数据库会话建立以后，Oracle Net 负责客户端和服务器端的数据传递。

图 3-1 显示了普通的 C/S 结构程序客户端和服务器端如何通过 Oracle Net 进行连接。

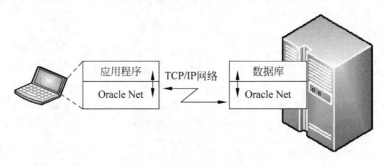

图 3-1　客户端/数据库连接

Java 应用程序通过 JDBC（Java Database Connectivity）驱动来连接 Oracle 数据库，Oracle 提供以下两类 JDBC 驱动。

1. JDBC OCI Driver

使用这类驱动，需要应用软件运行机器上安装 Oracle 客户端软件。应用程序通过 JDBC 驱动发请求，JDBC 驱动再通过 Oracle Net 将请求发送给服务器，如图 3-2 所示。

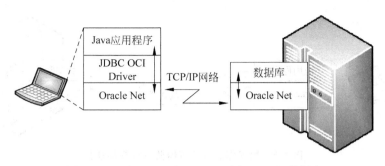

图 3-2 通过 JDBC OCI Driver 访问 Oracle 数据库

2. JDBC Thin Driver

这个是使用纯 Java 技术实现的驱动,应用软件机器上不需安装 Oracle 客户端软件,JDBC 驱动通过调用 Java Net(Java 技术实现的 Oracle Net)与数据库通信,如图 3-3 所示。

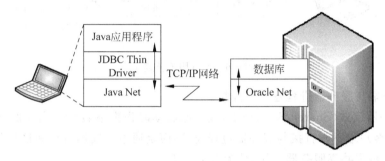

图 3-3 通过 JDBC Thin Driver 访问 Oracle 数据库

Oracle Net 包含两个软件组件,如图 3-4 所示。

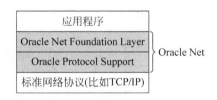

图 3-4 Oracle Net 组件

应用程序通过 Oracle 网络基础层(Oracle Net Foundation Layer)建立和维护与数据库服务器的连接,Oracle 网络基础层使用 Oracle 协议支持层(Oracle Protocol Support)与标准的网络协议通信,支持标准协议有 TCP/IP、命名管道(Named Pipes)等。

3.1.2 Oracle Net Listener

Oracle Net Listener 是人们所熟知的监听器,Oracle 数据库服务器通过监听器接收来自客户端的连接,图 3-5 显示了监听器接收客户端连接请求的过程。

专有服务器模式(Dedicated Server)下,监听器为每个客户端连接启动一个独立的专有服务器进程为其服务,会话结束后,服务器进程终止。图 3-6 显示了专有服务器连接,分为以下三步。

(1) 监听器接收一个客户端连接请求。

(2) 监听器启动一个服务器进程,并将客户连接请求转给服务器进程。

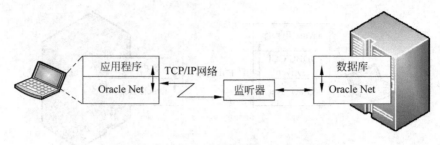

图 3-5　数据库通过监听器监听客户端连接

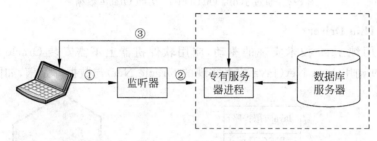

图 3-6　专有服务器连接

（3）客户端与服务器进程建立了连接。

共享服务器模式（Shared Server）下，数据库启动时根据参数设置启动服务器进程以及分发器进程，客户端不再直接与服务器进程连接，而是通过分发器进程间接与服务器进程通信。图 3-7 显示了共享服务器连接，分为以下三步。

（1）监听器接收一个客户端连接请求。
（2）监听器将连接请求转给分发器进程。
（3）客户端与分发器进程建立了连接。

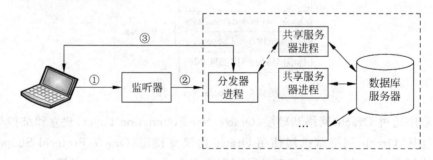

图 3-7　共享服务器连接

3.2　数据库服务注册

Oracle 客户端程序连接数据库时，需要先连接到负责监听客户端连接请求的监听器，这个连接需要监听器的 IP 地址和监听端口信息，一个监听器可能同时为多个数据库服务，客户端还要指定要连接的数据库信息，这时监听器才能在指定的数据库实例启动服务器进程来为客户端服务。Oracle 数据库包含实例部分和数据库部分，数据库启动后会自动将数据库信息注册到本地监听器,数据库名注册为服务名,服务名关联到实例名,客户端连接数据库需要数据库注册后的服务名,监听器根据服务名在服务名关联的数据库实例启动服务器进程。

3.2.1 服务的概念

Oracle 数据库作为一种服务展现给客户端,数据库默认的服务名是数据库名称,也可以通过参数配置多个服务名。图 3-8 显示了两个数据库,每个数据库都有单独的实例,不同的客户端通过服务名连接到不同的数据库实例。

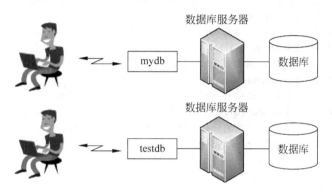

图 3-8　通过不同服务名连接不同数据库

一个数据库通常具有一个服务名,默认就是数据库名。可以通过系统参数(SERVICES_NAMES)配置多个不同的服务名,也可以使用包 DBMS_SERVICE 创建服务名。配置多个服务名,使不同的应用程序连接数据库时使用不同的服务名,Oracle 数据库内部有针对服务的统计信息,方便对不同应用的监控,另外也可以使用 Oracle 数据库中的资源管理器针对不同服务合理分配系统资源。图 3-9 显示两个客户端通过不同服务名访问同一个数据库。

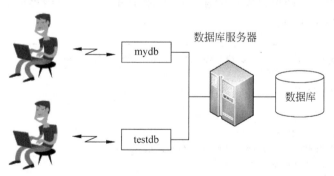

图 3-9　通过不同服务名连接同一数据库

【例】　查看监听器中数据库服务,数据库实例名 mysid,数据库名 mydb。

```
$ lsnrctl status
… …
Services Summary...
Service "mydb" has 1 instance(s).
  Instance "mysid", status READY, has 1 handler(s) for this service...
The command completed successfully
```

【例】 配置多个服务名。

```
$ sqlplus / as sysdba
SQL> alter system set service_names = srv1,srv2,srv3;
System altered.

$ lsnrctl status
…
Services Summary...
Service "SRV1" has 1 instance(s).
  Instance "mysid", status READY, has 1 handler(s) for this service...
Service "SRV2" has 1 instance(s).
  Instance "mysid", status READY, has 1 handler(s) for this service...
Service "SRV3" has 1 instance(s).
  Instance "mysid", status READY, has 1 handler(s) for this service...
Service "mydb" has 1 instance(s).
  Instance "mysid", status READY, has 1 handler(s) for this service...
The command completed successfully
```

3.2.2 服务注册

服务注册是指 Oracle 数据库在监听器中注册数据库信息,客户端向监听器发连接数据库请求时,如果提供的服务名尚未在监听器中注册,则连接失败。Oracle 数据库对外提供服务前必须向监听器进行服务注册,服务注册有以下两种形式。

1. 静态注册

静态注册指在监听器配置文件中通过脚本配置监听器要监听的数据库服务,监听器启动后配置的数据库服务即可供客户端连接使用,但此时并不知道数据库实例是否启动。

【例】 配置静态注册。

```
$ vi /u01/app/product/11.2.0/dbhome_1/network/admin/listener.ora
增加如下内容:
SID_LIST_LISTENER =
  (SID_LIST =
    (SID_DESC =
      (GLOBAL_DBNAME = test)
      (ORACLE_HOME = /u01/app/product/11.2.0/dbhome_1)
      (SID_NAME = mysid)
    )
  )

$ lsnrctl status
…
Services Summary...
Service "mydb" has 1 instance(s).
  Instance "mysid", status READY, has 1 handler(s) for this service...
Service "test" has 1 instance(s).
  Instance "mysid", status UNKNOWN, has 1 handler(s) for this service...
The command completed successfully
```

上例中通过在监听器配置文件 Listener.ora 中增加 SID_LIST_LISTENER 配置项来配置要监听的数据库信息,其中 GLOBAL_DBNAME 项配置的是静态注册的服务名,这样在监听器状态查询中可看到服务名 test,且对应的实例 mysid 状态为 UNKNOWN。

2. 动态注册

动态注册是指数据库启动后主动向监听器发送注册信息,数据库关闭注册信息清除。当数据库启动时,后台进程 PMON 负责向监听器注册,根据数据库实例启动所在的机器名(select host_name from v$instance)以及监听器的默认监听端口 1521 发起连接,如果连接成功,则发送数据库信息给监听器,如果连接失败,则每隔 1min 再次尝试连接。

如果监听器没有在默认端口 1521 监听,则 PMON 根据默认信息将无法完成动态注册,需要通过修改系统参数 LOCAL_LISTENER 指定监听器配置信息,这时 PMON 方能完成注册。

【例】 配置监听器在非 1521 端口监听,修改系统参数完成动态注册。

```
$ vi /u01/app/product/11.2.0/dbhome_1/network/admin/listener.ora
将端口修改为 1522
LISTENER =
  (DESCRIPTION =
    (ADDRESS = (PROTOCOL = TCP)(HOST = oracle11g)(PORT = 1522))
  )

修改参数 local_listener 指定监听器信息
$ sqlplus / as sysdba
SQL > alter system set local_listener =
'(DESCRIPTION = (ADDRESS = (PROTOCOL = tcp)(HOST = oracle11g)(PORT = 1522)))';
```

3.3 监听器配置

3.3.1 监听器配置文件

监听器的配置文件是 listener.ora,文件默认存储在 $ORACLE_HOME/network/admin 下,也可以通过环境变量 TNS_ADMIN 指定其他存储路径,监听器的配置信息主要是协议、主机名或 IP 地址、监听端口。

如果没有配置监听器的配置文件,则所有配置参数采用默认值,可以启停监听器,这时监听器默认名称是 LISTENER。监听器文件主要配置参数说明见表 3-1。

表 3-1 listener.ora 参数说明

配置参数	参数说明	范　例
ADDRESS	指定监听器地址,在参数 DESCRIPTION 中使用	ADDRESS= (PROTOCOL=tcp)(HOST=pc123)(PORT=1521) PROTOCOL 指定协议,HOST 指定主机名或 IP 地址,PORT 指定监听端口号

续表

配置参数	参数说明	范 例

DESCRIPTION	包含监听器地址	LISTENER =
		(DESCRIPTION=
		  (ADDRESS=(PROTOCOL=tcp)(HOST=pc123)
		     (PORT=1521)
		  )
		)

QUEUESIZE	指定并发连接请求数量	LISTENER =
		(DESCRIPTION=
		  (ADDRESS=(PROTOCOL=tcp)(HOST=pc123)
		     (PORT=1521)(QUEUESIZE=20)
		  )
		)

RECV_BUF_SIZE	指定会话接收缓冲区大小	LISTENER =
		(DESCRIPTION=
		  (ADDRESS=(PROTOCOL=tcp)(HOST=pc123)
		     (PORT=1521)(RECV_BUF_SIZE=11784)
		  )
		)

SEND_BUF_SIZE	指定会话发送数据缓冲区大小	LISTENER =
		(DESCRIPTION=
		  (ADDRESS=(PROTOCOL=tcp)(HOST=pc123)
		     (PORT=1521)(SEND_BUF_SIZE=11784)
		  )
		)

CONNECTION_RATE_Listener_name	指定连接数率,每秒连接数	CONNECTION_RATE_LISTENER=10

RATE_LIMIT	指定是否开启连接数率限制	LISTENER=
		(ADDRESS_LIST=
		  (ADDRESS=(PROTOCOL=tcp)(HOST=pc123)
		     (PORT=1521)(RATE_LIMIT=yes)
		  )
		)

ADR_BASE_listener_name	指定诊断信息存储路径	ADR_BASE_LISTENER = /u01/app
TRACE_LEVEL_listener_name	指定跟踪信息级别	TRACE_LEVEL_listener=admin

续表

配置参数	参数说明	范例

SID_LIST_
listener_name
	配置静态注册	SID_LIST_LISTENER =
		(SID_LIST =
		  (SID_DESC =
		    (GLOBAL_DBNAME = test)
		    (ORACLE_HOME=
		      /u01/app/product/11.2.0/dbhome_1)
		    (SID_NAME = mysid)
		  )
		)
		
		SID_LIST 指静态监听的实例列表
		SID_DESC 指实例描述
		GLOBAL_NAME 指定静态注册服务名，自定义
		ORACLE_HOME 指 Oracle 数据库软件安装目录
		SID_NAME 指定静态注册的数据库实例名

【例】 listener.ora 文件示例。

```
SID_LIST_LISTENER =
  (SID_LIST =
    (SID_DESC =
      (GLOBAL_DBNAME = test)
      (ORACLE_HOME = /u01/app/product/11.2.0/dbhome_1)
      (SID_NAME = mysid)
    )
  )

LISTENER =
  (DESCRIPTION =
    (ADDRESS = (PROTOCOL = TCP)(HOST = oracle11g)(PORT = 1521))
  )
ADR_BASE_LISTENER = /u01/app
```

3.3.2 监听器配置与管理

1. 使用 NETMGR 配置监听器

配置监听器可以使用配置工具 NETCA 或 NETMGR，也可以直接编辑监听器配置文件。NETCA 的使用方法参见 1.3.1 节。使用 NETMGR 配置监听器：

```
$ netmgr
```

命令执行后，出现如图 3-10 所示的配置界面，选择 Oracle Net Configuration | Local | Listeners，然后单击＋按钮。

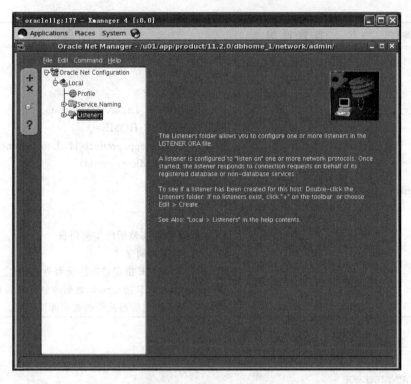

图 3-10　NETMGR 主页面

在图 3-11 显示监听器名称设置页面,默认监听器名为 LISTENER,单击 OK 按钮。

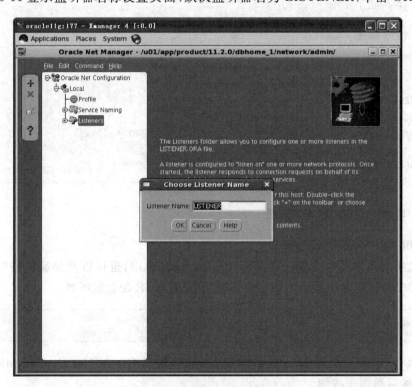

图 3-11　监听器名称设置页面

在如图 3-12 所示页面中，选择 Oracle Net Configuration | Local | Listeners | LISTENER，在右侧窗口显示监听器设置页面，单击 Add Address 按钮，显示监听信息设置页面，配置协议、主机名和监听端口，接下来如图 3-13 所示选择 File | Save Network Configuration 命令保存配置信息。

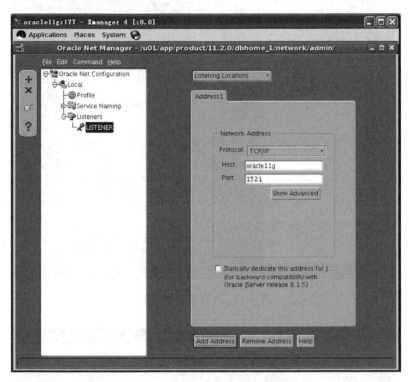

图 3-12　监听器地址设置页面

配置结束查看配置文件 listener.ora，验证配置信息是否正确。

2. 配置数据库静态注册

配置数据库静态注册使用 NETMGR 或直接修改监听器配置文件，下面演示 NETMGR 配置静态注册方法。

```
$ netmgr
```

命令执行后，出现如图 3-14 所示的界面，选择 Oracle Net Configuration | Local | Listeners | LISTENER。在如图 3-15 所示页面中选择 Database Services。

在如图 3-16 所示页面中输入静态注册信息，Global Databas Name 处输入静态注册的服务名，Oracle Home Directory 处输入 Oracle 数据库软件安装目录，SID 处输入静态注册服务关联的数据库实例名，接下来如图 3-17 所示选择 File | Save Network Configuration 命令保存配置信息。

3. 监听器控制工具 LSNRCTL

LSNRCTL 是监听器的命令控制工具，可用该工具完成基本的监听器启动停止、状态查询等。命令语法格式如下：

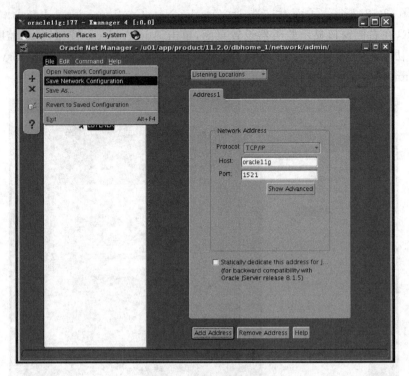

图 3-13　保存监听器配置

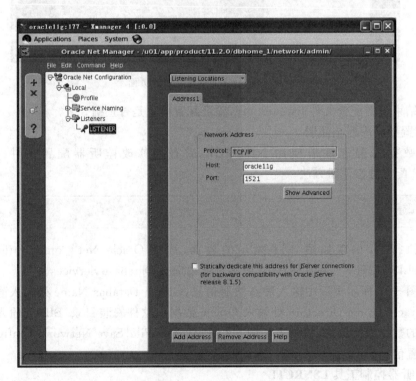

图 3-14　NETMGR 主页面

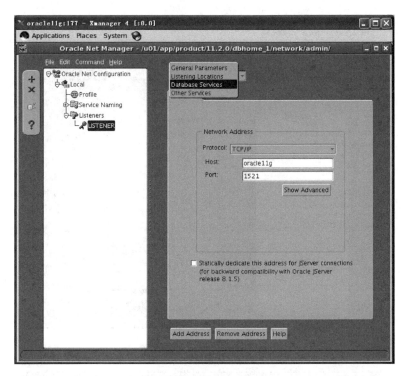

图 3-15 监听器设置页面

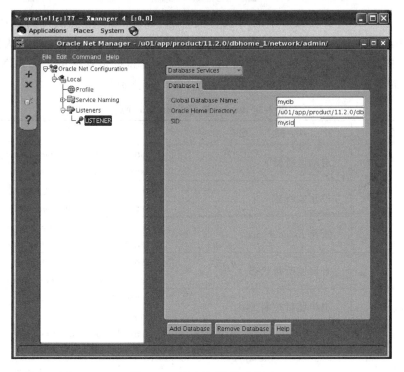

图 3-16 静态注册配置页面

Oracle Database 11g 管理与开发

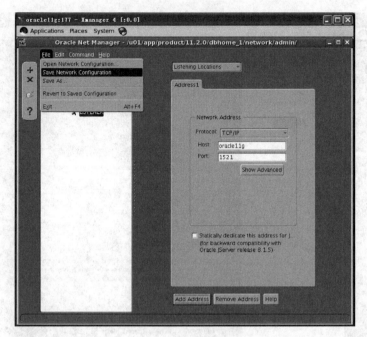

图 3-17　保存监听器配置

```
lsnrctl command listener_name
```

其中,lsnrctl 是 Oracle 安装软件中的可执行文件名,command 指要执行的具体命令,比如 start 或 stop,listener_name 指用命令控制的监听器名,如果名称省略,则默认控制名称为 LISTENER 的监听器。在配置监听器时,如果监听器名称不是默认的 LISTENER,则使用该控制工具时务必指定监听器名称。常用 command 见表 3-2。

表 3-2　lsnrctl 常用 command 介绍

命　　令	描　　述	范　　例
start	启动监听器	lsnrctl start [ ]listener_name] lsnrctl>start [listener_name]
stop	停止监听器	lsnrctl stop [listener_name] lsnrctl>stop [listener_name]
reload	重读 listener.ora	lsnrctl reload [listener_name] lsnrctl>reload [listener_name]
status	查询监听器状态	lsnrctl status [listener_name] lsnrctl>status[listener_name]
quit 或 exit	退出监听控制工具	lsnrctl>quit lsnrctl>exit
show parameter	显示当前监听器设置参数	lsnrctl show trc_level lsnrctl>show trc_level
trace level	设置跟踪	lsnrctl trace admin [listener_name] lsnrctl>trace admin [listener_name]

3.4 客户端连接

3.4.1 配置本地命名解析

客户端连接远程数据库时需要监听器的协议、IP 地址、监听端口号以及注册的数据库服务名,这些信息可以直接写在连接命令上,比如:

```
sqlplus u1/u1@192.168.1.2:1521/mydb
```

也可以将这些信息配置成一个网络服务命名,连接时只需在@符后写网络服务命名,比如:

```
sqlplus u1/u1@remote
```

Oracle 数据库客户端使用的命名解析方法有以下几种。

(1) 本地命名解析(Local Naming)。使用本地磁盘文件进行名称解析,每个客户端需单独配置。将网络服务命名和对应的连接信息存储到命名解析文件中,这个文件是 tnsnames.ora,默认存储在 $ORACLE_HOME/network/admin 下。

(2) 目录命名解析(Directory Naming)。使用目录服务器进行解析,需要配置目录服务器,进行集中解析,减少每个客户端配置工作量。

(3) 简单连接命名解析(Easy Connect Naming)。连接数据时直接写连接数据库信息,不需要其他解析文件或服务器。

(4) 外部命名解析(External Naming)。使用第三方的命名解析服务。

在这 4 个解析方法中,最常用的是本地命名解析。可以使用工具 NETCA、NETMGR 对文件 tnsnames.ora 进行配置,也可以直接使用文本编辑器进行文件修改。NETCA 的使用方法参见 1.5.2 节,接下来介绍使用 NETMGR 配置 tnsnames.ora。

此处客户端环境为 Windows XP 系统,在"运行"对话框中输入"netmgr",如图 3-18 所示,或是选择【开始】|【程序】| Oracle - OraClient11g_home1【配置和移植工具】| Net Manager。

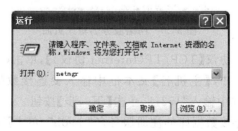

图 3-18 "运行"对话框

在图 3-19 中选择【服务命名】,单击＋按钮,弹出如图 3-20 所示的网络服务名输入页面,输入"remote",单击【下一步】按钮。

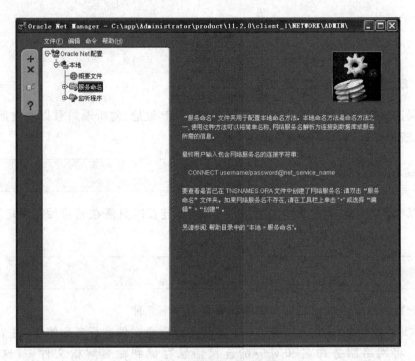

图 3-19　选择【服务命名】

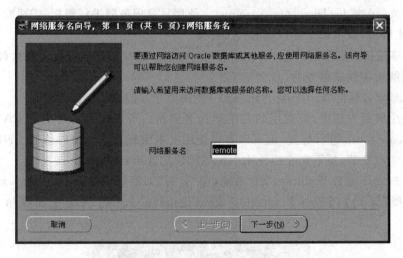

图 3-20　网络服务名输入页面

在图 3-21 中选择网络协议【TCP/TP(Internet 协议)】,单击【下一步】按钮。

在如图 3-22 所示页面中,【主机名】文本框中输入要连接数据库所在机器的 IP"192.168.1.2",【端口号】文本框中输入"1521",单击【下一步】按钮。

在图 3-23 中输入要连接数据库的注册服务名"mydb",单击【下一步】按钮。

在图 3-24 中可单击【测试】按钮,验证基本信息是否正确。此处单击【测试】按钮。

图 3-25 显示测试成功页面,如果不成功,可通过单击【更改登录】按钮修改测试用的用户名和口令,然后再单击【测试】按钮。此处测试成功,单击【关闭】按钮。然后在如图 3-26

图 3-21 选择网络协议

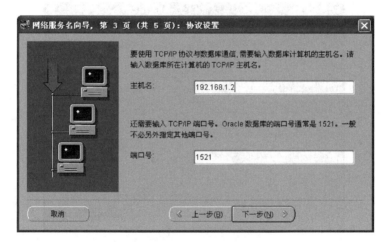

图 3-22 IP 和端口设置

图 3-23 数据库服务名设置

所示页面中选择【文件】|【保存网络配置】命令保存配置信息。

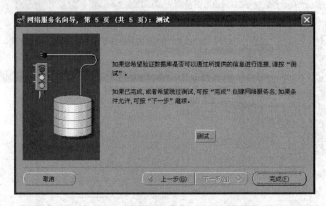

图 3-24 测试页面

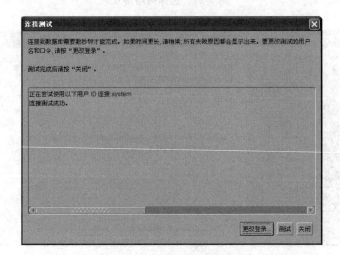

图 3-25 测试成功页面

图 3-26 保存网络服务名配置

配置结束后，检查本地命名解析文件内容如下：

```
REMOTE =
  (DESCRIPTION =
    (ADDRESS_LIST =
      (ADDRESS = (PROTOCOL = TCP)(HOST = 192.168.1.2)(PORT = 1521))
    )
    (CONNECT_DATA =
      (SERVICE_NAME = mydb)
    )
  )
```

验证登录是否成功：

```
C:\> sqlplus system/oracle@remote
SQL*Plus: Release 11.2.0.1.0 Production on Fri Jul 5 11:02:00 2013
Copyright (c) 1982, 2010, Oracle.   All rights reserved.


Connected:
Oracle Database 11g Enterprise Edition Release 11.2.0.1.0 - Production
With the Partitioning, OLAP, Data Mining and Real Application Testing option
SQL>
```

3.4.2 本地命名解析文件

本地命名解析文件 tnsnames.ora 是包含网络服务名映射到连接数据库信息的配置文件，默认情况下存储在 $ORACLE_HOME/network/admin 下，也可以通过系统环境变量 TNS_ADMIN 指定到其他存储目录。文件内容基本格式如下：

```
net_service_name =
(DESCRIPTION =
   (ADDRESS = (protocol_address_information))
   (CONNECT_DATA =
     (SERVICE_NAME = service_name)
   )
)
```

其中，net_service_name 指定网络服务名，自定义名称；protocol_address_information 部分配置监听器的连接信息，包括协议、主机名或 IP 地址、监听端口；service_name 部分配置要连接的数据库服务名。命名解析文件 tnsnames.ora 主要配置参数说明见表 3-3。

表 3-3　tnsnames.ora 参数说明

配置参数	参数说明	范　　例
DESCRIPTION	网络服务命名的描述开始	网络服务名＝ (DESCRIPTION＝ 　(ADDRESS＝…) 　(CONNECT_DATA＝…))
ADDRESS	指定监听器地址	网络服务名＝ (DESCRIPTION＝ 　(ADDRESS＝(PROTOCOL＝tcp) 　　　　　(HOST＝pc123)(PORT＝1521)) 　(CONNECT_DATA＝…))
CONNECT_DATA	指定要连接的数据库服务名	网络服务名＝ (DESCRIPTION＝ 　(ADDRESS＝(PROTOCOL＝tcp) 　　　　　(HOST＝pc123)(PORT＝1521)) 　**(CONNECT_DATA＝(SERVICE_NAME＝mydb))**)
LOAD_BALANCE	设置客户端是否启用负载均衡,如设置为 yes,则当有多个监听地址时,可随机选择一个进行连接,均衡监听器的负载	网络服务名＝ (DESCRIPTION＝ 　**(LOAD_BALANCE＝yes)** 　(ADDRESS＝(PROTOCOL＝tcp) 　　　　　(HOST＝pc001)(PORT＝1521)) 　(ADDRESS＝(PROTOCOL＝tcp) 　　　　　(HOST＝pc002)(PORT＝1521)) 　(CONNECT_DATA＝(SERVICE_NAME＝mydb)))
FAILOVER	设置连接失败的切换,当有多个监听器时,如果连接失败,则依次尝试连接每个监听器,如果没有设置均衡负载,则从第一个开始连接	网络服务名＝ (DESCRIPTION＝ 　**(LOAD_BALANCE＝yes)** 　**(FAILOVER＝yes)** 　(ADDRESS＝(PROTOCOL＝tcp) 　　　　　(HOST＝pc001)(PORT＝1521)) 　(ADDRESS＝(PROTOCOL＝tcp) 　　　　　(HOST＝pc002)(PORT＝1521)) 　(CONNECT_DATA＝(SERVICE_NAME＝mydb)))

续表

配置参数	参数说明	范例
CONNECT _TIMEOUT	设置连接超时时间，单位为秒	网络服务名＝ (DESCRIPTION＝ 　(**CONNECT_TIMEOUT＝10**) 　(ADDRESS＝(PROTOCOL＝tcp) 　　　　　　(HOST＝pc001)(PORT＝1521)) 　(CONNECT_DATA＝(SERVICE_NAME＝mydb)))
RETRY_COUNT	尝试连接次数，默认为0	网络服务名＝ (DESCRIPTION＝ 　(**CONNECT_TIMEOUT＝10**) 　(**RETRY_COUNT＝2**) 　(ADDRESS＝(PROTOCOL＝tcp) 　　　　　　(HOST＝pc001)(PORT＝1521)) 　(CONNECT_DATA＝(SERVICE_NAME＝mydb)))

小　　结

　　Oracle 数据库客户端和服务器进程间传递数据有特定的协议，这部分实现称为 Oracle Net。本章分为 4 个部分进行了讲解：一是介绍 Oracle 网络服务组件；二是讲解数据库服务注册概念，分别介绍了静态注册和动态注册；三是讲解监听器配置，包括监听器配置文件讲解、配置工具讲解以及监听器控制工具使用；四是讲解客户端配置，包括客户端连接数据库方式、本地命名解析文件作用和配置方法。

思　考　题

　　1. 如何静态注册数据库？写出具体配置信息。
　　2. 何为动态注册数据库？简述注册过程。
　　3. 监听器配置文件是什么？有几种配置方法？
　　4. 假设监听器采用 TCP，主机名为 home，端口是 1522，监听器名为 LISTENER，写出在配置文件中的配置信息。
　　5. 本地命名解析文件什么？举例说明文件内容格式。
　　6. 假设连接信息如下：协议 TCP，对方主机 IP 192.168.1.1，端口 1555，服务名为 orcl，配置网络服务命名为 test，写出配置信息。

第 4 章　逻辑存储结构

了解数据在数据库中的存储组织方式是深入学习 Oracle 数据库的必经之路。本章将通过数据库逻辑结构的介绍，逐步为读者展现 Oracle 数据库内部的数据存储方式。

4.1　逻 辑 结 构

4.1.1　逻辑结构简介

数据库逻辑存储结构包括表空间、段、区、块 4 个逻辑概念：块是 Oracle 数据库数据的基本存储单位；区由一组连续的块组成，作为段空间分配的单位；段是指数据段，是存储数据的空间，数据库中有存储空间要求的对象都有对应的数据段，比如表和索引；表空间是指存储数据段的空间，对应到磁盘上的数据文件。

图 4-1 展现了数据库逻辑概念之间的关系以及与物理文件结构的对应关系。其中，Tablespace 是表空间，Segment 是段，Extent 是区，Oracle Data Block 是数据库块，Data File 是数据文件，OS Block 是操作系统块。

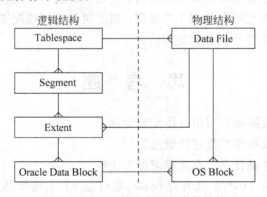

图 4-1　Oracle 数据库逻辑存储结构

一个表空间可包含多个段，一个段只能属于一个表空间；一个段可包含多个区，一个区只属于一个段，一个段包含的多个区可以分别属于不同的数据文件；每个区是由一组连续的数据块组成；一个表空间对应多个数据文件，但一个数据文件只能属于一个表空间；一个数据文件包含多个区，区不能跨数据文件，只能属于一个数据文件；一个数据库块包含多个操作系统块。图 4-2 显示了一个表空间内部段、区、块之间的关系，假设一个段占用空间 128KB，共有两个区，这两个区分别在数据文件 1 和数据文件 2 上，每个区占空间 64KB，由

8个连续的8KB大小的数据库块组成。

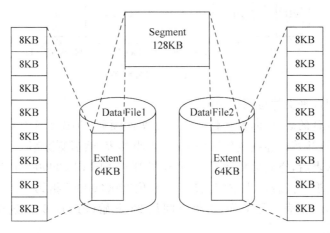

图4-2 段、区、块之间关系

4.1.2 区管理方式

区是给段分配空间的逻辑单位，由一组连续的块组成。当数据库中多个对象都需要空间时，区的分配和回收方式将直接决定分配和回收效率。

1. 字典管理表空间

字典管理表空间（Dictionary Managed Tablespace，DMT）使用数据字典表来管理区的分配。当一个区被分配出去或是一个区被回收都需要更新数据字典表。比如，当一个表需要分配区，则首先查询空闲区的数据字典表（fet＄，空闲区的字典表 free extent），如果有空闲空间可用分配一组块，然后修改空闲区字典表，同时在已分配区的字典表中增加一条（uet＄，已分配区的字典表 used extent），这种情况下，数据库依赖对字典表的修改来管理区的分配。

通过修改数据库字典表来分配区，这种后台执行的修改工作称为递归SQL（recursive SQL），频繁地递归SQL将会对数据库造成一些负面影响，在并发的情况下，这种修改需要顺序执行，性能也不高。当前Oracle 11g中所有表空间的区管理方式都是本地管理，已经不建议使用字典管理，而且如果SYSTEM表空间是本地管理方式则不允许创建字典管理表空间，如果通过SQL创建数据库指定SYSTEM表空间为字典管理，则其他表空间可以是字典管理或本地管理。

2. 本地管理表空间

本地管理表空间（Locally Managed Tablespace，LMT）在表空间的数据文件头中通过位图的方式记录区的分配情况，与字典管理表空间的区别在于，字典管理表空间属于集中式管理，通过两个字典表管理所有表空间的区分配，而本地管理表空间属于分散管理，每个数据文件头只记录该文件内部的区分配情况，并发性远远高于字典管理方式。

本地管理表空间在数据文件头维护一个位图跟踪空闲的和使用的空间，当空间分配或释放时，Oracle数据库通过修改位（bit）的状态来反映空间分配情况。1代表分配的空间，0代表空闲的空间，每位代表64KB空间，位图方式管理空间如图4-3所示。

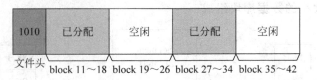

图 4-3 文件空间分配位图

【例】 表空间数据文件位图记录测试。

创建一个 50MB 的表空间,默认情况下数据文件预留 1MB(128 个数据块)空间作文件头,数据段从 128 块开始。连续创建三张 2MB 大小的表,然后删除第二张表,观察位图记录信息。删除第二张表后,空间释放用 0 代表,这样"FFFFFFFF00000000 FFFFFFFF"中前 8 位"FFFFFFFF"指第一张表占用空间,接下来 8 位"00000000"指第二张表释放空间,最后"FFFFFFFF"指第三张表占用空间,可用空间从 32 开始(First:32),这里 32 指的是第 32 个 64KB 大小的区。

```
SQL > create tablespace tbs datafile 'c:\tbs.dbf' size 50M;
SQL > create table t01(id number) tablespace tbs storage(initial 2m);
SQL > create table t02(id number) tablespace tbs storage(initial 2m);
SQL > create table t03(id number) tablespace tbs storage(initial 2m);
SQL > drop table t02;

跟踪文件头信息如下(截取):
...
File Space Bitmap Block:
BitMap Control:
RelFno: 7, BeginBlock: 128, Flag: 0, First: 32, Free: 63424
FFFFFFFF00000000 FFFFFFFF00000000 0000000000000000 0000000000000000
0000000000000000 0000000000000000 0000000000000000 0000000000000000
...
```

4.1.3 段空间管理方式

段空间管理指管理分配给段的数据块,记录段内数据块的使用情况,当用户需要插入新数据时,数据库根据块的使用情况决定使用哪个块。

1. 自动段空间管理

自动段空间管理(Automatic Segment Space Management,ASSM)使用位图记录数据块的使用状态,每个数据块的使用状态分为 FS1、FS2、FS3、FS4 和 FULL,其中 FS1 代表 0~25% 空闲空间,FS2 代表 25%~50% 空闲空间,FS3 代表 50%~75% 空闲空间,FS4 代表 75%~100% 空闲空间,FULL 代表块已满。

假设每个数据块的状态用 4 个比特位代表,0001 代表 FS1,0010 代表 FS2,0011 代表 FS3,0100 代表 FS4,0101 代表 FULL,每组块中的第一块作为位图块,则段空间位图管理方式如图 4-4 所示。

图 4-4 中的位图块是一级位图块,一级位图块之上还有二级位图块,如果数据量很大,可能会有三级、四级甚至更高级别的位图块。图 4-5 为一个两级位图示例,在自动段空间管

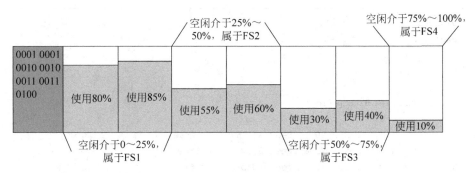

图 4-4 位图管理段空间

理方式下,当多个会话同时需要插入数据时,都需要找到可以插入的块,首先通过段头找到二级位图块,再根据二级位图中的一级位图地址分散查找不同的一级位图,在每个一级位图下查找可以插入数据的数据块。

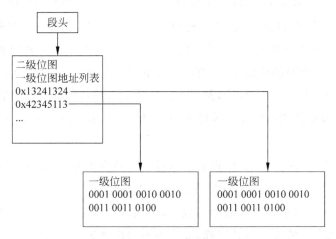

图 4-5 位图层次

【例】 查看位图块中记录的数据块信息,二级位图中记录的是一级位图的地址,一级位图中记录数据块的使用状态,其中每个分配区的前几个标识为"Metadata"的块为一级位图块,其余为数据块,记录每个数据块使用状态。

```
二级位图
Dump of Second Level Bitmap Block
...
  L1 Ranges :
  --------------------------------------------------
  0x01c00180   Free: 1 Inst: 1
  0x01c00190   Free: 1 Inst: 1
  0x01c001a0   Free: 1 Inst: 1
  0x01c001b0   Free: 1 Inst: 1
  0x01c001c0   Free: 1 Inst: 1
  0x01c001d0   Free: 1 Inst: 1
  0x01c001e0   Free: 1 Inst: 1
```

```
        0x01c001f0   Free: 1 Inst: 1
        0x01c00200   Free: 1 Inst: 1
        0x01c00201   Free: 5 Inst: 1
...
一级位图
Dump of First Level Bitmap Block
...
   0:Metadata       1:25-50% free    2: 25-50% free
   4:50-75% free    5:50-75% free    6:50-75% free
   8:50-75% free    9:50-75% free   10:50-75% free
  12:FULL          13: FULL         14: FULL
...
```

2. 手动段空间管理

手动段空间管理（Manual Segment Space Management，MSSM）使用 FREE LIST 管理块的使用。FREE LIST（自由链表）是一种单向链表，用于定位可以接收数据的块，字典管理表空间和本地管理表空间中都可以使用，Oracle 使用自由链表来管理可用的数据块。Oracle 记录了有空闲空间的块用于 INSERT，空闲空间来源于两种方式：①段中所有超过 HWM（High Water Mark，高水位线，是段中已用块和未用块的分界线）的块，这些块已经分配给段了，但是还未被使用；②段中所有在 HWM 下的且链入了 FREE LIST 的块，可以被重用。

当需要插入新数据时，首先根据进程号选择使用哪条自由链表（假如有多个链表，则使用 PID%NFL+1，PID 为进程号，NFL 为自由链表数量，即进程号对自由链表数量取余加1），然后在自由链表中获取可插入数据的数据块，当块不能满足插入新数据要求时，数据库认为该块已满并从自由链表中将其移除，如果所有自由链表中的块都不能满足要求，则水位线后移并格式化新块，将新块添加到自由链表，直到找到可用块为止。当从已满的数据块中删除数据时，数据占用空间较低时将重新添加到自由链表。实现过程主要由以下几个属性控制。

PCTFREE：指定块预留空闲空间，为数据库修改（UPDATE）所用，默认 10%，当块数据达到 100%-PCTFREE 时，该块不再插入新数据，从自由链表中移除。

PCTUSED：指定块数据使用百分比，当块数据满之后，从自由链表移除，当块数据删除直到低于这个百分比方可重新添加到自由链表做插入新数据用。

FREELISTS：指定每组内自由链表的数量。当多个会话并发访问时，可通过增加自由链表数量提升性能。

FREELIST GROUPS：指定自由链表组的数量，每个自由链表组在段头占用单独的块，在并发量很大时，增大自由链表组，可缓解链表所在数据块的访问压力。每个段创建时都创建一个主链表（Master Free List），为缓解链表压力通常创建多个进程链表（Process Free List），由主链表协调管理，如图 4-6 所示。

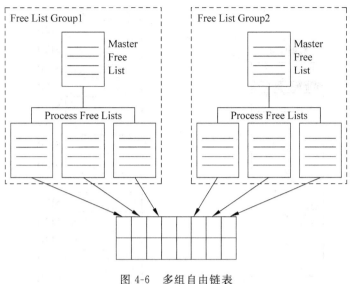

图 4-6 多组自由链表

4.2 数 据 块

4.2.1 数据块结构

数据块(Oracle Data Blocks)是 Oracle 数据库中最小的存储单位,数据都存放在数据块中。一个块占用一定的磁盘空间,可以是 2KB、4KB、8KB、16KB 或 32KB。每次请求数据的时候,都是以块为单位,比如一个查询语句只查询表中的一行数据,这时服务器进程会将该行数据所在的数据块从磁盘文件中读出放入数据缓冲池,然后再从数据缓冲池中的块搜索数据。

数据库中默认的块标准大小由初始化参数 DB_BLOCK_SIZE 指定,具有标准大小的块称为标准块(Standard Block),块的大小和标准块的大小不同的块叫非标准块(Nonstandard Block)。同一数据库中,Oracle 可以同时支持多种块大小,针对不同块大小设置相对应的数据缓冲区。

数据库块和操作系统块不同,数据库块通常都是操作系统块的整数倍,是一组操作系统块。每次读磁盘数据时,数据库通过操作系统执行 I/O,而操作系统每次执行 I/O 的时候,是以操作系统的块为单位,Oracle 每次执行 I/O 的时候,都是以 Oracle 的数据块为单位。Oracle 数据块和操作系统块对应关系参见图 4-7。

Oracle 数据库在数据块中存储数据时有特定的内部格式,包括头部信息、空闲空间和数据,如图 4-8 所示。

(1) 块头(Common and Variable):存放块的基本信息,如块的物理地址、块所属的段的类型(是数据段还是索引段),以及事务信息。

(2) 表目录(Table Directory):存放表的信息,即如果一些表的数据被存放在这个块中,那么,这些表的相关信息将被存放在"表目录"中,一般情况下块中只存储一个表的数据,使用簇时块中会有多个表数据。

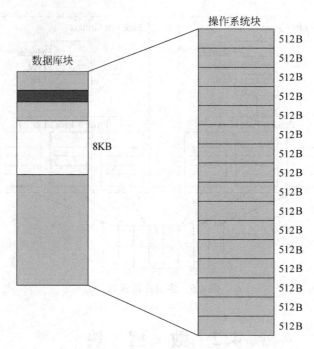

图 4-7　数据库块与操作系统块对应关系

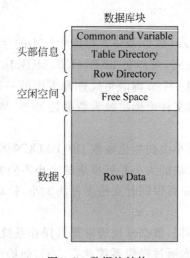

图 4-8　数据块结构

（3）行目录（Row Directory）：如果块中有行数据存在，则这些行的信息将被记录在行目录中。这些信息包括行的地址等。行目录信息不会随着行数删除而释放，这些空间将被重复使用。

（4）空闲空间（Free Space）：空闲空间是一个块中未使用的区域，这片区域用于新行的插入和已经存在的行的更新。

（5）行数据（Row Data）：是真正存放表数据和索引数据的地方，这部分空间是已被数据行占用的空间。

头部信息包含块头（Common and Variable）、表目录（Table Directory）和行目录（Row

Directory),头部信息区不存放数据,它存放整个块的信息,头部信息区的大小是可变的。

4.2.2 行数据存储格式

数据块中的行数据部分是真正包含实际数据的区域,比如表数据和索引数据,每个数据行也像数据块一样有特定的格式,行数据是变长的,一个行可能包含一个或多个行片段(Row Piece),每个行片段都有行头(Row Header)和列数据(Column Data)。行格式参见图4-9。

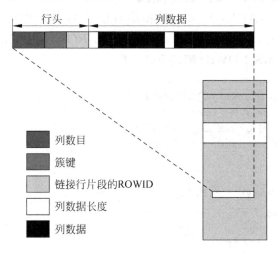

图 4-9 行数据格式

1. 行结构信息

1) 行头

Oracle数据库使用行头来管理存储在块中的行片段,行头包含以下几个信息项。

(1) 列数目:指定该行片段中存储的列数据的数量。

(2) 链接行片段信息:如果一行数据能够插入到单个块中,则一行数据只包含一个行片段,否则会有多个行片段出现,这些行片段通过ROWID进行连接。

(3) 簇键:当创建的是簇表时,此处存储簇键值。

2) 列数据

在行头后面,行中列的数据依次排列,数据长度+实际数据,列数据的存储顺序基本与建表时列的顺序一致,如果有LONG类型数据,则LONG类型放在最后。

2. ROWID

Oracle数据库使用ROWID(行号)唯一标识一行数据,ROWID中包含访问一行数据的信息,ROWID不在数据库中直接存储。扩展的ROWID中包括数据对象编号、行所在文件编号、行所在数据块编号以及在数据块上行的顺序号,使用六十四进制编码(A~Z代表0~25,a~z代表26~51,0~9代表52~61,+代表62,/代表63)。

【例】 查看表中数据的行编号。

```
SQL> select rowid from t01;
ROWID
------------------
AAAEoeAAFAAAAAOAAA
```

行编号的格式如图 4-10 所示。

图 4-10 ROWID 格式

按照上例中的查询结果可知：数据对象编号"AAAEoe"为 4×64×64+40×64+30＝18 974；所在文件编号"AAF"为 5；所在数据块编号"AAAAAO"为 14；所在块上的行顺序号"AAA"为 0，即第一行；也可通过 Oracle 包 DBMS_ROWID 中的函数求得。

【例】 通过包 DBMS_ROWID 解析 ROWID。

```
SQL> select dbms_rowid.rowid_object(rowid) object_no,
  dbms_rowid.rowid_relative_fno(rowid) file_no,
  dbms_rowid.rowid_block_number(rowid) block_no,
  dbms_rowid.rowid_row_number(rowid) row_no
  from t01;

OBJECT_NO      FILE_NO      BLOCK_NO      ROW_NO
----------     --------     ---------     ------
    18974         5             14           0
```

4.3 区

区是 Oracle 数据库给段分配空间的逻辑单元，段的存储空间按区来管理，区由一组逻辑上连续的数据库块组成。一个段随着数据的增长，可以获得更多的区空间。当创建一个段时，会分配一个初始区，初始区默认使用的是段所在表空间的初始区的大小。在 Oracle 11g 中可以通过使用参数 deferred_segment_creation 来设置创建数据库对象时是否延迟创建段，默认是 true，也就是创建数据库对象时不创建段，在第一次插入数据需要存储空间时创建段。

【例】 创建数据库对象，验证段创建。

```
SQL> create table t01(id number);

SQL> select extents,blocks from user_segments where segment_name = 'T01';
no rows selected.

SQL> insert into t01 values(100);

SQL> select extents,blocks from user_segments where segment_name = 'T01';
   EXTENTS      BLOCKS
   -------      ------
      1           8
```

每个段都有段头数据块，段头数据块包含所有的区列表，区列表中列出区的起始数据块地址和该区的数据块总数,除此以外还包括每个区所在的一级位图块和数据块列表,见下面示例。

【例】 查看段头数据块记录信息。

```
...
Extent Map
-----------------------------------------------------------------
  0x01000088   length: 8
  0x01000090   length: 8
  0x01000098   length: 8

Auxillary Map
-----------------------------------------------------------------
  Extent 0   :  L1 dba:  0x01000088  Data dba:  0x0100008b
  Extent 1   :  L1 dba:  0x01000088  Data dba:  0x01000090
  Extent 2   :  L1 dba:  0x01000098  Data dba:  0x01000099
-----------------------------------------------------------------

Second Level Bitmap block DBAs
-----------------------------------------------------------------
  DBA 1:   0x01000089
```

上例中显示该数据库对象存储数据的段包含三个区,第一个区起始块地址为 0x01000088,共 8 块,第二、三个区地址为 0x01000090 和 0x01000098,也都包含 8 个块。第一个区所在一级位图块地址为 0x01000088,起始存储数据的块为 0x0100008b;第二区所在一级位图块地址也是 0x01000088,起始存储数据的块为 0x01000090;第三个区所在一级位图块地址为 0x01000098,起始存储数据的块为 0x01000099;这个段只有一个二级位图块,地址是 0x01000089。

字典管理表空间中区大小的分配可以通过存储参数来指定,本地管理表空间中区大小的分配方式有以下两种。

(1) 统一大小(Uniform Size)。创建表空间时通过参数 Uniform Size 指定一个区大小,则该表空间中所有对象分配区时按照统一标准进行分配。统一大小可提高区的使用效率,避免碎片产生,但如果区定义较小对于大表分配可能过于频繁。

(2) 自动分配(Autoallocation)。自动分配方式由 Oracle 数据库自行根据表数据的量进行动态调整,基本原则是大表分大区,小表分小区。分配基本规则是:表数据<8MB,每个区大小 64KB;表数据<64MB 且>8MB,每个区大小 1MB;表数据<1GB 且>64MB,每个区大小 8MB;表数据>1GB,每个区大小 64MB。

4.4 段

段是指数据段,指数据的存储空间,由一个或多个区组成。在 Oracle 数据库中有数据存储空间要求的常见用户段类型有表、表分区、簇、索引、索引分区以及大对象(LOB),除了用户段还有几种特殊的段,包括临时段和 undo 段。

4.4.1 用户段创建

默认情况下,当创建表或索引时 Oracle 11g 使用延迟段创建,当用户插入数据时数据库将为表、索引以及大对象创建段。数据库参数 deferred_segment_creation 可用来控制是否延迟段创建。图 4-11 显示了用户段的创建。

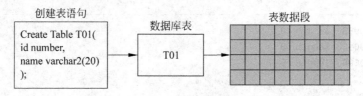

图 4-11 数据库表和表数据段创建

如果创建表包含主键约束或唯一约束,则数据库自动为主键约束或唯一约束创建唯一索引,如果表中包含 LOB 列数据,数据库将为 LOB 列单独创建段以及 LOB 列的索引段。这样如下面的建表语句将会创建 4 个段,如图 4-12 所示。

```
create table lob(
id number primary key,
lobv clob
);
```

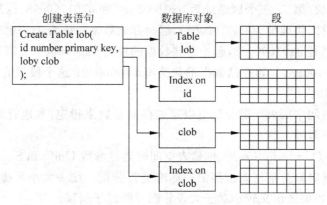

图 4-12 一个表对应多个段

4.4.2 临时段和 undo 段

1. 临时段

临时段在数据库用户的临时表空间中创建。当处理一个大数据量的排序查询,服务器进程内存空间往往满足不了排序需要,这时会借助临时段来存储数据,还有哈希连接、位图索引的创建与合并等都需要临时段,临时段操作完成后由数据库自动清除。

除了上述应用外,数据库为临时表和索引也分配临时段,临时表的数据只保留到事务结束或会话结束,且多个会话在使用同一临时表时,数据库为每个会话单独分配区,每个会话只能访问分给这个会话的区,多个会话间临时表的数据相互独立。

2. undo 段

Oracle 数据库每开启一个事务,都需要分配一个 undo 段,这个 undo 段用来记录 undo 数据,当用户执行 rollback 命令时,数据库读取事务对应的 undo 段中数据回滚事务。除了用于回滚事务外,undo 段记录的数据还用于一致性读、实例恢复和闪回操作。

4.5 表空间

4.5.1 表空间介绍

表空间是所有数据段的逻辑存储空间,表空间将数据存储到一个或多个数据文件或临时文件。创建数据库默认表空间如图 4-13 所示。

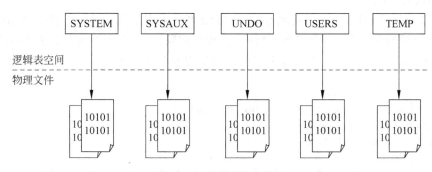

图 4-13 系统默认表空间

1. SYSTEM 表空间

SYSTEM 表空间是数据库创建必须包含的表空间,Oracle 数据库使用 SYSTEM 表空间来管理数据库。SYSTEM 表空间包含数据字典表、管理用的表和视图以及编译的存储过程、函数、触发器,这些对象都是 SYS 用户拥有。

用 DBCA 创建数据库时,SYSTEM 表空间默认是本地管理方式,这种情况数据库内所有表空间都必须是本地管理。

2. SYSAUX 表空间

SYSAUX 表空间是 SYSTEM 表空间的辅助表空间,用于存储 SYSTEM 表空间没有存储的数据库元数据,比如企业管理器(Enterprise Manager)和数据库流(Oracle Stream)组件的数据。这个表空间在创建数据库时自动创建。

3. UNDO 表空间

Oracle 数据库在自动 UNDO 管理模式下使用 UNDO 表空间保存数据库产生的 UNDO 数据,数据库可以存在多个 UNDO 表空间,但只有一个是系统当前使用的,通过参数 undo_tablespace 指定。

4. USERS 表空间

USERS 表空间是创建数据库时创建的一个普通用户表空间,可供普通用户建表使用。

5. TEMP 表空间

TEMP 表空间是一个临时表空间,用于在用户会话期间存储一些临时数据,比如内存排序空间不足时使用 TEMP 表空间,或者是临时表存储数据用。

表空间从数据存储类型可分为持久表空间(Permanent Tablespace)、临时表空间(Temporary Tablespace)和UNDO表空间(Undo Tablespace)。持久表空间用于长期存储用户数据,比如表和索引数据,上述表空间中SYSTEM表空间、SYSAUX表空间和USERS表空间都是持久表空间。临时表空间用于临时存储数据,比如排序的中间数据或临时表的数据。UNDO表空间用于存储UNDO数据。

表空间从使用状态上可分为只读表空间(Read Only)、可读可写表空间(Read Write),从表空间是否在线可分为在线表空间(Online)和离线表空间(Offline)。表空间正常状态下基本都是在线且可读可写。

4.5.2 表空间基本操作

1. 创建表空间

使用Create Tablespace创建表空间,可以创建持久表空间、UNDO表空间和临时表空间,创建表空间通常由具有SYSDBA权限或DBA权限用户来创建,普通用户如果被授予了Create Tablespace权限则也可创建表空间。创建表空间时数据库一定是open状态。

语法:

```
CREATE  [ SMALLFILE | BIGFILE ]  [ TEMPORARY | UNDO ]
TABLESPACE tablespace_name
{ DATAFILE | TEMPFILE } 'filename'
[ SIZE n {M | G | T }]
[ AUTOEXTEND {OFF| ON [NEXT n {M | G | T } ] [MAXSIZE {UNLIMITED | n {M | G | T }}]  }
[BLOCKSIZE nK ]
[ LOGGING | NOLOGGING ]
[ NO FORCE LOGGING | FORCE LOGGING ]
[ ONLINE | OFFLINE ]
[ EXTENT MANAGEMENT LOCAL { AUTOALLOCATE | UNIFORM [ SIZE n {M | G | T }]} ]
[ SEGMENT SPACE MANAGEMENT { AUTO | MANUAL } ]
```

属性说明:

1) SMALLFILE | BIGFILE

smallfile指小文件表空间,是传统的Oracle表空间文件类型,也是默认选项,一个表空间的数据文件或临时文件最多可达1022个文件,每个文件最多可包含400万(2^{22})个数据块,如果块的大小为8KB,则单个文件最大可达32GB,表空间最大可达32TB,如果块大小为32KB,则单个文件最大可达128GB,表空间最大可达128TB。

bigfile指大文件表空间,一个表空间只包含一个文件,文件最多可包含40亿(2^{32})数据块,如果块的大小为8KB,则单个文件最大可达32TB,如果块大小为32KB,则单个文件最大可达128TB。

大文件表空间文件少管理简单,但文件移动困难且受文件系统最大文件限制,小文件表空间文件虽多,但移动较容易。

2) TEMPORARY | UNDO

temporary是创建临时表空间的关键字,undo是创建UNDO表空间的关键字,默认不写则创建的是普通持久表空间。

3) tablespace_name

tablespace_name 指定符合 Oracle 数据库命名规则的表空间名称。

4) DATAFILE | TEMPFILE

DATAFILE | TEMPFILEdatafile 用于指定创建持久表空间或 UNDO 表空间对应的数据文件，tempfile 用于指定创建临时表空间的临时文件。

5) filename

filename 指定表空间对应的数据文件名，如果使用 OMF（Oracle Managed File），通过参数指定文件存储路径（db_create_file_dest），直接写文件名即可，如果没有使用 OMF，可以指定全路径文件名。

6) SIZE

SIZE 指定文件的初始大小，如未指定默认为 100MB。

7) AUTOEXTEND OFF|ON

autoextend 用于指定文件是否可自动扩展，autoextend off 指禁止自动扩展，autoextend on 指启用自动扩展，默认情况下设置自动扩展。当设置自动扩展时，一定要通过 next 语句设置每次扩展大小，如果没有设置，默认每次扩展一个数据块。自动扩展可减少手动扩展或添加数据文件的操作，但要时刻监测磁盘空间使用状态。

8) MAXSIZE

maxsize 指定数据文件最大可扩展的大小，unlimited 指没有限制，或者通过指定大小来限制。

9) BLOCKSIZE nK

用 blocksize 指定非标准块大小的表空间，可以是 2K、4K、8K、16K、32K，如果创建表空间时没有指定块大小，则采用数据库标准块大小（db_block_size）。在创建非标准块大小表空间时，必须事先配置相应块大小的缓冲区（db_nK_cache_size）。另外，临时表空间不可以指定非标准块大小。

10) LOGGING | NOLOGGING

指定表空间内数据库对象（表、表分区等）的默认日志属性，logging 记日志，nologging 不记日志，logging 是默认的。临时表空间和 UNDO 表空间此属性无效。如果在创建表时指定日志属性，则表空间级别的日志属性不起作用。

11) NO FORCE LOGGING | FORCE LOGGING

force logging 是强制日志属性，表空间内的所有对象强制记录日志，即使创建表时指定 nologging。默认情况下不是强制日志模式。

12) ONLINE | OFFLINE

指定表空间创建时是 online 或是 offline，online 状态指表空间创建后立即可用，是默认选项，offline 指定表空间创建后不可用，直到状态修改为 online 后才可使用。

13) EXTENT MANAGEMENT LOCAL { AUTOALLOCATE | UNIFORM }

extent management local 指定表空间为本地管理，autoallocate 指定表空间中区的分配是数据库自动分配，按照大表分大区小表分小区的规则进行，参见 4.3 节，uniform 指定区的分配是统一大小，默认为 1MB。默认为 extent management local autoallocate。

14) SEGMENT SPACE MANAGEMENT { AUTO | MANUAL }

segment space management 指定段空间管理方式，auto 指使用位图方式自动管理，manual 指使用自由链表来管理，参见 4.1.3 节。默认为 segment space management auto。

【例】 创建持久表空间，包含一个文件，自动增长。

```
SQL > create tablespace tbs1
    datafile 'tbs1.dbf' size 10M autoextend on next 1M;
```

【例】 创建持久表空间，包含两个文件，自动增长。

```
SQL > create tablespace tbs2
    datafile
    'tbs201.dbf' size 10M autoextend on next 1M,
    'tbs202.dbf' size 10M autoextend on next 1M;
```

【例】 创建持久表空间，包含一个文件，自动增长，块大小 16KB。

```
SQL > alter system set db_16k_cache_size = 5M;
SQL > create tablespace tbs3
    datafile 'tbs3.dbf' size 10M autoextend on next 1M
    blocksize 16k;
```

【例】 创建临时表空间，包含一个文件，自动增长。

```
SQL > create temporary tablespace tmp1
    tempfile 'tmp1.dbf' size 10M autoextend on next 1M;
```

【例】 创建 UNDO 表空间，包含一个文件，自动增长。

```
SQL > create undo tablespace undotbs
    datafile 'undotbs.dbf' size 10M autoextend on next 1M;
```

2. 修改表空间

使用 Alter Tablespace 修改表空间，可以增加或删除数据文件或临时文件，或修改表空间属性，比如日志属性等。修改表空间通常由具有 SYSDBA 权限或 DBA 权限的用户来创建，普通用户如果被授予了 Alter Tablespace 权限则也可修改表空间。

语法：

```
ALTER TABLESPACE tablespace_name
[ ADD { DATAFILE | TEMPFILE } file_specification
 | DROP {DATAFILE | TEMPFILE } { filename | fileno }
 | RENAME DATAFILE old_filename TO new_filename ]
[ LOGGING | NOLOGGING | FORCE LOGGING | NO FORCE LOGGING]
[ ONLINE | OFFLINE {NORMAL | TEMPORARY | IMMEDIATE} ]
[ READ ONLY | READ WRITE ]
```

```
[ AUTOEXTEND ON [NEXT nM] [MAXSIZE {UNLIMITED| nM}] | off ]
[ RESIZE nM ]
[ RETENTION { GUARANTEE | NOGUARANTEE } ]
[ RENAME TO newname ]
[ {BEGIN | EDD} BACKUP ]
```

属性说明：

1）tablespace_name

tablespace_name 指定要修改的表空间名。

2）ADD { DATAFILE | TEMPFILE }

ADD { DATAFILE | TEMPFILE }为表空间添加文件，持久表空间添加数据文件，临时表空间添加临时文件，文件名和自动扩展属性同创建文件中的设置。

3）DROP {DATAFILE | TEMPFILE}

根据文件名或编号删除空的数据文件或临时文件，文件删除规则有：必须是空的，不能包含任何段；不能是表空间的第一个文件；必须是在线状态。

4）RENAME DATAFILE

RENAME DATAFILE 可以修改表空间的数据文件名，但首先必须先置表空间为 offline。

5）LOGGING | NOLOGGING | FORCE LOGGING | NO FORCE LOGGING

LOGGING | NOLOGGING | FORCE LOGGING | NO FORCE LOGGING 修改表空间的日志属性。表空间修改成 logging 或 nologging，将影响修改后创建的对象，之前创建的对象日志属性不受影响。

6）ONLINE | OFFLINE {NORMAL | TEMPORARY | IMMEDIATE}

ONLINE | OFFLINE {NORMAL | TEMPORARY | IMMEDIATE}指定 online 可以使表空间在线，指定 offline 可以使表空间离线。指定离线时，如果是 normal 方式，数据库将数据缓冲区中所有属于该表空间数据文件的数据块写出到数据文件，normal 是默认方式，重新在线时不需要恢复；如果是 temporary 方式，数据库将数据缓冲区中所有属于该表空间的在线数据文件的数据块写出到数据文件，重新在线时，原来离线状态的文件需要恢复；如果是 immediate 方式，则马上离线，不写数据文件，但重新在线时需要恢复。

7）READ ONLY | READ WRITE

READ ONLY | READ WRITE 指定 read only 将表空间置于只读状态，任何针对该表空间上对象的 DML 语句不被允许，恢复成 read write 模式后，方可进行 DML 操作。

8）AUTOEXTEND

AUTOEXTEND 指定 autoextend 语句修改大文件表空间的自动扩展属性，小文件表空间不适用。小文件表空间修改自动扩展属性使用 alter database 语句。

9）RESIZE

RESIZE 指定 resize 语句修改大文件表空间的文件大小，小文件表空间不适用。小文件表空间修改文件大小使用 alter database 语句。

10）RETENTION { GUARANTEE | NOGUARANTEE }

RETENTION { GUARANTEE | NOGUARANTEE }指定 retention 语句用于修改 UNDO 表空间保留的 UNDO 数据在事务结束后的保留策略，guarantee 是强制保留到指定时间(undo_retention)，只有超过该时间范围占用空间才可被重用，noguarantee 不强制保

留，占用空间在空间不足时可被重用。

11) REMAME

REMAME 指定 rename 用于修改表空间名称。

12) {BEGIN | EDD} BACKUP

在数据库归档模式下，开始或结束表空间的热备份。

【例】 修改表空间，增加数据文件。

```
SQL> create tablespace tbs
    datafile 'tbs01.dbf' size 10M autoextend on next 1M;

SQL> alter tablespace tbs
    add datafile 'tbs02.dbf' size 10M autoextend on next 1M;
```

【例】 修改文件大小。

```
SQL> alter database datafile 'tbs02.dbf' resize 100M;
```

【例】 离线表空间。

```
SQL> alter tablespace tbs offline;
```

【例】 将表空间恢复在线，并修改成只读。

```
SQL> alter tablespace tbs online;
SQL> alter tablespace tbs read only;
```

3. 删除表空间

使用 Drop Tablespace 删除表空间，删除表空间通常由具有 SYSDBA 权限或 DBA 权限的用户来删除，普通用户如果被授予了 Drop Tablespace 权限也可删除表空间。表空间无论是在线还是离线均可删除。

语法：

```
DROP TABLESPACE tablespace_name
[ INCLUDING CONTENTS [ AND DATAFILES] [CASCADE CONSTRAINTS]]
```

属性说明：

1) tablespace_name

tablespace_name 指定要删除的表空间名。

2) INCLUDING CONTENTS

当表空间不空，包含数据库对象时，必须指定该语句，否则删除失败。

3) AND DATAFILES

在语句 including contents 后指定 and datafiles，使数据库在删除表空间同时，删除操作系统上的文件，如果是 OMF 方式不用指定该语句。

4) CASCADE CONSTRAINTS

如果其他表空间中的表有通过外键引用该表空间中的表时,需要指定该语句,否则删除失败。

【例】 删除表空间。

```
SQL> drop tablespace tbs including contents and datafiles;
```

4.6 相关视图

1. DBA_TABLESPACES

DBA_TABLESPACES 视图显示数据库中所有表空间信息,参见表 4-1。

表 4-1 DBA_TABLESPACES 主要列介绍

列　　名	描　　述
TABLESPACE_NAME	表空间名
BLOCK_SIZE	表空间数据块大小,默认是标准数据块,8KB
INITIAL_EXTENT	初始区大小
NEXT_EXTENT	下个区大小
STATUS	表空间状态 ONLINE、OFFLINE、READ ONLY
CONTENTS	表空间存储内容类型 UNDO、PERMANENT、TEMPORARY
LOGGING	表空间默认日志属性 LOGGING、NOLOGGING
FORCE_LOGGING	表空间是否强制记日志,YES 或 NO
EXTENT_MANAGEMENT	区管理方式 LOCAL、DICTIONARY
ALLOCATION_TYPE	区分配方式 SYSTEM、UNIFORM、USER
RETENTION	UNDO 信息保留策略 GUARANTEE、NOGUARANTEE
SEGMENT_SPACE_MANAGEMENT	段空间管理方式 AUTO、MANUAL
BIGFILE	是否大文件表空间,YES 或 NO

2. DBA_FREE_SPACE

DBA_FREE_SPACE 显示数据库中所有表空间的空闲空间信息,参见表 4-2。

表 4-2 DBA_FREE_SPACE 主要列介绍

列　　名	描　　述
TABLESPACE_NAME	表空间名
FILE_ID	文件绝对编号

续表

列　名	描　述
BLOCK_ID	空闲空间起始块号
BYTES	字节数
BLOCKS	块数
RELATIVE_FNO	文件相对编号

3. DBA_DATA_FILES

DBA_DATA_FILES 显示数据库中所有数据文件信息,参见表 4-3。

表 4-3　DBA_DATA_FILES 主要列介绍

列　名	描　述
FILE_NAME	文件名
FILE_ID	文件绝对编号
TABLESPACE_NAME	所属表空间名
BYTES	字节数
BLOCKS	块数
STATUS	状态,AVAILABLE 或 INVALID
RELATIVE_FNO	文件相对编号
AUTOEXTENSIBLE	是否可扩展,YES 或 NO
MAXBYTES	最大字节数
MAXBLOCKS	最多块数
INCREMENT_BY	每次扩展块数
USER_BYTES	用户可用字节数
USER_BLOCKS	用户可用块数

4. DBA_EXTENTS

DBA_EXTENTS 显示数据库中所有区信息,具体列参见表 4-4。

表 4-4　DBA_EXTENTS 主要列介绍

列　名	描　述
OWNER	所属用户
SEGMENT_NAME	段名
SEGMENT_TYPE	段类型,比如 TABLE、INDEX
TABLESPACE_NAME	表空间名
EXTENT_ID	区编号
FILE_ID	所在文件号
BLOCK_ID	区起始块号
BYTES	字节数
BLOCKS	块数
RELATIVE_FNO	相对文件编号

5. DBA_SEGMENTS

DBA_SEGMENTS 显示数据库中所有段信息,具体列参见表 4-5。

表 4-5　DBA_SEGMENTS 主要列介绍

列　　名	描　　述
OWNER	所属用户
SEGMENT_NAME	段名
SEGMENT_TYPE	段类型，比如 TABLE、INDEX
TABLESPACE_NAME	表空间名
HEADER_FILE	段头所在文件号
HEADER_BLOCK	段头所在数据块
BYTES	字节数
BLOCKS	块数
EXTENTS	区数
INITIAL_EXTENT	初始区大小
NEXT_EXTENT	下个区大小
MIN_EXTENTS	最少区个数
MAX_EXTENTS	最多区个数
MAX_SIZE	最多的块数
FREELISTS	自由链表数
FREELIST_GROUPS	自由链表组数
RELATIVE_FNO	相对文件号
BUFFER_POOL	存储缓冲池 DEFAULT、KEEP、RECYCLE

小　　结

　　Oracle 数据库逻辑结构主要讲解空间的分配管理以及数据的存储。本章首先讲解表空间、段、区、数据库块之间的关系，然后针对这 4 个部分分别进行了讲解。数据块是 Oracle 读写数据的基本单位；区是一组连续的数据库块，是给段分配空间的单位，区的管理方式包括 DMT 和 LMT；段是指一段存储空间，由一个或多个区组成，有数据存储空间要求的对象都有段，段内的空间使用管理方式包括 ASSM 和 MSSM；表空间是一个逻辑概念，由一个或多个数据文件组成。

思　考　题

　　1. 什么是 DMT？什么是 LMT？DMT 和 LMT 的区别是什么？现在主要用的是哪个？
　　2. 什么是 ASSM？什么是 MSSM？ASSM 和 MSSM 的区别是什么？
　　3. 手动段空间管理中，参数 PCTUSED、PCTFREE、FREE LIST 都是什么含义？
　　4. 段是什么？举例说明数据库中常见的段。
　　5. 表空间有哪些类型？简述其作用。
　　6. 创建表空间 tbs1，包含两个数据文件 tbs101.dbf 和 tbs102.dbf，初始大小都是 50MB，自动扩展，每次扩 10MB，写出创建语句。

7. 创建 UNDO 表空间 undo1，数据文件 undo101.dbf，初始大小 100MB，自动扩展，每次扩 10MB，写出创建语句。

8. 创建临时表空间 tmp1，数据文件 tmp1.dbf，初始大小 100MB，自动扩展，每次扩 10MB，写出创建语句。

9. 查询所有表空间大小、使用空间大小、使用百分比、剩余空间大小、剩余百分比。

第 5 章　用户权限管理

数据安全一直是企业最关心的话题,数据库作为专业的数据存储技术更加关心数据安全。本章将从用户管理、权限控制和安全审计三个方面讲解 Oracle 数据库的安全体系。

5.1　用户管理

数据库用户是指有权限访问数据库的用户,用户信息由数据库来管理。一个应用要访问数据库,需要使用正确的用户名和口令连接数据库。每个 Oracle 数据库创建时都会创建一些用户,比如 SYSTEM 用户、SYS 用户,SYSTEM 用户是 Oracle 数据库具有 DBA 角色的用户,SYS 用户是 Oracle 数据库中具有 SYSDBA 权限的用户,也是权限最大的用户。

5.1.1　创建用户

Oracle 数据库中通常使用 SYS 用户或 SYSTEM 用户来创建用户,如果希望普通用户能够创建用户,需要授予该用户 CREATE USER 权限。新创建的用户默认权限为空,登录 Oracle 数据库,至少需要授予 CREATE SESSION 系统权限。

创建用户语法如下:

```
CREATE USER user
IDENTIFIED { BY password | EXTERNALLY | GLOBALLY }
[ DEFAULT TABLESPACE tablespace_name ]
[ TEMPORARY TABLESPACE tablespace_name ]
[ QUOTA { UNLIMITED | size } ON tablespace_name ]
[ PROFILE profile ]
[ PASSWORD EXPIRE ]
[ ACCOUNT { LOCK | UNLOCK } ]
[ ENABLE EDITIONS ];
```

属性说明:

1. user

指定要创建的用户名称。

2. IDENTIFIED

指定 Oracle 数据库验证用户的方式:

BY password 指定创建一个本地用户且需要口令方可登录数据库,口令区分大小写;

EXTERNALLY 指定创建一个外部验证用户,可以是操作系统验证或第三方软件验证;

GLOBALLY 指定创建一个全局用户,可以通过企业目录服务验证。

这三种验证方式中,BY password 最为常用。

3. DEFAULT TABLESPACE

指定用户创建数据库对象默认使用的表空间。假如未指定,则使用数据库的默认表空间,如果数据库没有指定默认表空间,则使用 SYSTEM 表空间。

```
SQL> select property_value from database_properties
    where property_name = 'DEFAULT_PERMANENT_TABLESPACE';
PROPERTY_VALUE
---------------------------------------
USERS
```

用户默认表空间不能是临时表空间和 UNDO 表空间。

4. TEMPORARY TABLESPACE

指定用户创建临时段时使用的临时表空间或临时表空间组。假如未指定,则使用数据库的默认临时表空间,如果数据库没有指定默认临时表空间,则使用 SYSTEM 表空间。

```
SQL> select property_value from database_properties
    where property_name = 'DEFAULT_TEMP_TABLESPACE';
PROPERTY_VALUE
---------------------------------------
TEMP
```

5. QUOTA

指定用户在表空间中可使用空间的最大值。如果指定 UNLIMITED 则意味着没有限制,只要表空间还有空闲就可以用,同一个用户可以在多个表空间中分配空间使用份额。如果在用户默认表空间中没有分配空间使用份额,则用户不能使用其空间。

6. PROFILE

给用户指定一个概要文件。概要文件可以限制用户的资源分配,如果没指定,则将概要文件 DEFAULT 指定给用户。

7. PASSWORD EXPIRE

指定用户口令过期。这个设置强制用户在第一次登录数据库时必须修改口令。

8. ACCOUNT

指定用户账号是否锁定。LOCK 锁定账号,用户不能登录;UNLOCK 不锁定账号,用户可登录,是系统默认设置。

9. ENABLE EDITIONS

它设置用户支持数据库对象的多版本。

【例】 创建用户 u01，口令 u01。

```
C:\> sqlplus system/oracle@remote
SQL> create user u01 identified by u01;
SQL> select default_tablespace, temporary_tablespace, profile
    from dba_users where username = 'U01';
DEFAULT_TABLESPACE      TEMPORARY_TABLESPACE      PROFILE
------------------      --------------------      -------
USERS                   TEMP                      DEFAULT
```

【例】 创建用户 u02，口令 u02，默认表空间 tbs1，临时表空间 tmp1。

```
C:\> sqlplus system/oracle@remote
SQL> create tablespace tbs1 datafile 'tbs1.dbf' size 50M;
SQL> create temporary tablespace tmp1 tempfile 'tmp1.dbf' size 50M;
SQL> create user u02 identified by u02
    default tablespace tbs1
    temporary tablespace tmp1;
SQL> select default_tablespace, temporary_tablespace, profile
    from dba_users where username = 'U02';
DEFAULT_TABLESPACE      TEMPORARY_TABLESPACE      PROFILE
------------------      --------------------      -------
TBS1                    TMP1                      DEFAULT
```

【例】 创建用户 u03，口令 u03，默认表空间 tbs1，临时表空间 tmp1，在表空间 tbs1 上可使用空间 20MB，在表空间 users 上使用无限制。

```
C:\> sqlplus system/oracle@remote
SQL> create user u03 identified by u03
    default tablespace tbs1
    temporary tablespace tmp1
    quota 20M on tbs1
    quota unlimited on users;
SQL> select tablespace_name, username, max_bytes from dba_ts_quotas
    where username = 'U03';
TABLESPACE_NAME      USERNAME      MAX_BYTES
---------------      --------      ---------
TBS1                 U03           20971520
USERS                U03           -1
```

【例】 创建用户 u04，口令 u04，默认表空间 tbs1，临时表空间 tmp1，在 tbs1 上使用无限制，口令初始过期。

```
C:\> sqlplus system/oracle@remote
SQL> create user u04 identified by u04
    default tablespace tbs1
    temporary tablespace tmp1
    quota unlimited on tbs1
    password expire;
SQL> select username,account_status,expiry_date from dba_users
    where username = 'U04';
USERNAME         ACCOUNT_STATUS    EXPIRY_DATE
----------       --------------    --------------
U04              EXPIRED           2013-7-10 12:29:38


C:\> sqlplus u04/u04@remote
SQL*Plus: Release 11.2.0.1.0 Production on Wed Jul 10 12:33:15 2013
Copyright (c) 1982, 2010, Oracle.  All rights reserved.

ERROR:
ORA-28001: the password has expired
Changing password for u01
New password:
```

【例】 创建用户 u05，口令 u05，默认表空间 tbs1，临时表空间 tmp1，在 tbs1 上使用无限制，账户初始锁定。

```
C:\> sqlplus system/oracle@remote
SQL> create user u05 identified by u05
    default tablespace tbs1
    temporary tablespace tmp1
    quota unlimited on tbs1
    account lock;
SQL> select username,account_status,expiry_date from dba_users
    where username = 'U05';
USERNAME         ACCOUNT_STATUS    EXPIRY_DATE
----------       --------------    --------------
U05              LOCKED            2013-7-10 12:29:38

C:\> sqlplus u05/u05@remote
SQL*Plus: Release 11.2.0.1.0 Production on Wed Jul 10 12:33:15 2013
Copyright (c) 1982, 2010, Oracle.  All rights reserved.

ERROR:
ORA-28000: the account is locked
Enter user-name:
```

【例】 创建用户 u06,口令 u06,默认表空间 tbs1,临时表空间 tmp1,在 tbs1 上使用无限制,支持多版本。

```
C:\> sqlplus system/oracle@remote
SQL> create user u06 identified by u06
    default tablespace tbs1
    temporary tablespace tmp1
    quota unlimited on tbs1
    enable editions;

C:\> sqlplus u06/u06@remote
SQL> create table t01 (id number,name varchar2(20));
SQL> insert into t01 values(1,'hello');
SQL> commit;

SQL> create edition tested;
SQL> create edition tested1 as child of tested;

SQL> alter session set edition = tested;
SQL> create editioning view v01 as select id from t01;
SQL> select * from v01;
    ID
------
     1
SQL> alter session set edition = tested1;
SQL> create or replace editioning view v01 as select name from t01;
SQL> select * from v01;
    ID
------
hello

SQL> select view_name,text,edition_name
    from user_views_ae where view_name = 'V01';
VIEW_NAME    TEXT                              EDITION_NAME
----------   -------------------------------   ------------
V01          select id from t01;               TESTED
V01          select name from t01;             TESTED1
```

5.1.2 修改用户

Oracle 数据库中通常使用 SYS 用户或 SYSTEM 用户来修改用户,如果希望普通用户能够修改用户,需要授予该用户 ALTER USER 权限。修改用户的大多数选项与创建用户选项含义一致,这里重点强调和角色、代理用户相关的选项。

修改用户语法如下:

```
ALTER USER user
IDENTIFIED { BY password | EXTERNALLY | GLOBALLY }
[ DEFAULT TABLESPACE tablespace ]
```

```
[ TEMPORARY TABLESPACE tablespace ]
[ QUOTA { UNLIMITED | size } ON tablespace ]
[ PROFILE profile ]
[ PASSWORD EXPIRE ]
[ ACCOUNT { LOCK | UNLOCK } ]
[ ENABLE EDITIONS ]
[ DEFAULT ROLE {role[, …]  |  ALL [ EXCEPT role[, …] ]  |  NONE ]
[ { GRANT | REVOKE } CONNECT THROUGH db_user_proxy ];
```

属性说明：

1) user

指定要修改的用户名称。

2) DEFAULT ROLE

指定用户的默认角色。用户被授予的角色初始都是默认角色，只要用户登录即可使用角色中的权限，如果希望不是所有角色登录即可使用，通过该语句进行设置。非默认的角色在用户登录后通过命令 SET ROLE 设置后即可使用。

3) { GRANT | REVOKE } CONNECT THROUGH

设置或取消通过代理用户登录。

【例】 创建用户 u1，授予两个角色 connect 和 resource，设置默认角色 connect，登录后验证。

```
C:\> sqlplus system/oracle@remote

创建用户
SQL> create user u1 identified by u1 quota unlimited on users;
User created.

授予角色
SQL> grant connect,resource to u1;
Grant succeeded.

修改默认角色
SQL> alter user u1 default role connect;
User altered.

登录查询默认角色
C:\> sqlplus u1/u1@remote
SQL> select * from session_roles;
ROLE
--------------------------------
CONNECT
```

【例】 创建用户 u2，将 u2 设置成用户 u1 的代理用户，并通过 u2 用户登录。

```
C:\> sqlplus system/oracle@remote

创建代理用户
SQL> create user u2 identified by u2;
User created.

将用户 u2 设置为 u1 的代理用户
SQL> alter user u1 grant connect through u2;
User altered.

通过 u2 完成 u1 用户登录
$ sqlplus u2[u1]/u2@remote
SQL> show user
USER is "U1"
```

5.1.3 概要文件

Oracle 通过概要文件(PROFILE)可以限制用户对系统资源的使用,以及设置用户口令相关参数来保证口令安全。概要文件常用参数介绍参见表 5-1。

表 5-1 概要文件常用属性

参数类型	参 数	描 述
系统资源	SESSIONS_PER_USER	指定每个用户最大并发的会话数
	CPU_PER_SESSION	指定每个会话的 CPU 时间限制,单位百分之一秒
	CPU_PER_CALL	指定每个调用的 CPU 时间限制,单位百分之一秒,比如一次解析、一次执行
	CONNECT_TIME	指定连接时间,超过后会话结束
	IDLE_TIME	指定连续的空闲时间,超过后会话结束
	LOGICAL_READS_PER_SESSION	指定每个会话读取数据块的限制
	LOGICAL_READS_PER_CALL	指定每个调用读取数据块的限制
	PRIVATE_SGA	指定每个会话可在 SGA 中分配的空间
	COMPOSITE_LIMIT	指定会话的资源成本
口令	FAILED_LOGIN_ATTEMPTS	指定最大失败尝试登录次数,超过后用户账号进入锁定状态
	PASSWORD_LIFE_TIME	指定口令最长使用天数
	PASSWORD_REUSE_TIME	指定口令重用间隔天数
	PASSWORD_REUSE_MAX	指定口令重用前修改次数
	PASSWORD_LOCK_TIME	指定失败尝试登录账户锁定天数
	PASSWORD_GRACE_TIME	指定口令到期后宽限天数
	PASSWORD_VERIFY_FUNCTION	指定口令格式验证函数

1. 创建概要文件

创建概要文件语法如下:

```
CREATE PROFILE profile LIMIT
param_name param_value [ … ] ;
```

属性说明：

1) profile

指定要创建的概要文件名称。

2) param_name param_value

指定资源参数或口令参数，多个参数设置用空格隔开。如果设置资源参数，则需要先设置系统参数 resource_limit。

【例】 创建一个概要文件，设置每个用户并发会话数 2，修改用户 u1 的概要文件。

```
C:\> sqlplus system/oracle@remote

设置资源限制参数
SQL> alter system set resource_limit = true;
System altered.

创建概要文件
SQL> create profile myprof1 limit sessions_per_user 2;
Profile created.

修改用户概要文件
SQL> alter user u1 profile myprof1;
User altered.

打开三个窗口，都以用户 u1 进行登录，第三个窗口登录信息：
$ sqlplus u1/u1@remote
ERROR:
ORA-02391: exceeded simultaneous SESSIONS_PER_USER limit
Enter user-name:
```

【例】 创建一个概要文件，设置失败尝试登录次数 2，用户口令有效期 30 天。

```
C:\> sqlplus system/oracle@remote

创建概要文件
SQL> create profile myprof2 limit
    failed_login_attempts 2
    password_life_time 30;
Profile created.

修改用户概要文件
SQL> alter user u1 profile myprof2;
User altered.

登录尝试，口令错误，第一次
C:\> sqlplus u1/123@remote
ERROR:
```

```
ORA - 01017: invalid username/password; logon denied

登录尝试,口令错误,第二次
C:\> sqlplus u1/123@remote
ERROR:
ORA - 01017: invalid username/password; logon denied

登录尝试,口令错误,第三次
C:\> sqlplus u1/123@remote
ERROR:
ORA - 28000: the account is locked
```

2. 修改概要文件

修改概要文件语法如下：

```
ALTER PROFILE profile LIMIT
param_name param_value [ … ] ;
```

属性说明：

1）profile

指定要创建修改的概要文件名称。

2）param_name param_value

指定资源参数或口令参数,多个参数设置用空格隔开。如果设置资源参数,则需要先设置系统参数 RESOURCE_LIMIT。

3. 删除概要文件

删除概要文件语法如下：

```
DROP PROFILE profile [ CASCADE ];
```

属性说明：

1）profile

指定要创建删除的概要文件名称。

2）CASCADE

删除概要文件同时取消与用户的关联。

4. 相关视图

DBA_PROFILES 视图显示数据库中所有概要文件信息,参见表 5-2。

表 5-2　DBA_PROFILES 列介绍

列　　名	描　　述
PROFILE	概要文件名
RESOURCE_NAME	资源参数名
RESOURCE_TYPE	资源参数类型
LIMIT	限制值

5.1.4 删除用户

删除用户语法如下：

```
DROP USER user [ CASCADE ];
```

属性说明：

1. user

指定要创建删除的用户名称。

2. CASCADE

删除用户前需要先删除用户的数据库对象。

5.1.5 相关视图

1. DBA_USERS

DBA_USERS 视图显示数据库中所有用户信息，参见表 5-3。

表 5-3 DBA_USERS 列介绍

列 名	描 述
USERNAME	用户名
ACCOUNT_STATUS	账号状态
DEFAULT_TABLESPACE	默认表空间
TEMPORARY_TABLESPACE	临时表空间
PROFILE	概要文件
EDITIONS_ENABLED	是否支持多版本

2. DBA_TS_QUOTAS

DBA_TS_QUOTAS 视图显示数据库所有用户的表空间可使用配额信息，参见表 5-4。

表 5-4 DBA_TS_QUOTAS 列介绍

列 名	描 述
TABLESPACE_NAME	表空间名
USERNAME	用户名
BYTES	已使用空间
MAX_BYTES	最大可使用空间

5.2 权限管理

数据库权限是用来限制用户保障数据安全的重要机制，用户如果不具备权限，则不能操作数据库的任何对象。数据库权限分为以下两类。

1. 系统权限

指用户执行特定类型 SQL 语句的权力,比如 CREAET TABLE、CREATE VIEW 等。

2. 对象权限

指访问其他用户数据库对象的权利,比如 SELECT ON u01.t01、UPDATE ON u01.t01 等。

5.2.1 系统权限

1. 权限授予

权限授予语法如下:

```
GRANT { system_privilege [, … ] | ALL PRIVILEGES } TO grantee
[ WITH ADMIN OPTION ];
```

属性说明:

1) system_privilege | ALL PRIVILEGES

指定要授予用户或角色的系统权限,可同时授予多个系统权限,权限用","隔开。也可以通过 ALL PRIVILEGES 将全部权限授予用户或角色。常用系统权限参见表 5-5。

2) grantee

指定被授权者,可以是用户名或角色名,可同时授予多个用户或角色。

3) WITH ADMIN OPTION

指定被授权者具有管理该权限的权力,被授权者可以将该权限授予其他用户或角色,也可以从其他用户或角色回收该权限。

表 5-5　常用系统权限

权　限	描　述
CREATE SESSION	连接数据库
CREATE TABLE	在用户 Schema 中创建表
ALTER TABLE	修改用户 Schema 中的表
DROP TABLE	删除用户 Scheme 中的表
CREATE CLUSTER	在用户 Schema 中创建簇
CREATE VIEW	在用户 Schema 中创建视图
CREATE MATERIALIZED VIEW	在用户 Schema 中创建物化视图
CREATE SEQUENCE	在用户 Schema 中创建序列
CREATE SYNONYM	在用户 Schema 中创建同义词
CREATE DATABASE LINK	在用户 Schema 中创建数据库链接
CREATE PROCEDURE	在用户 Schema 中创建过程、函数或包
CREATE TRIGGER	在用户 Schema 中创建触发器
SYSDBA	启停数据库、创建数据库、数据库修改、备份恢复
SYSOPER	启停数据库、数据库修改、备份恢复

【例】　创建用户 u1,授予 create session、create table 权限。

```
C:\> sqlplus system/oracle@remote

创建用户
SQL> create user u1 identified by u1 quota unlimited on users;
User created.

授予权限
SQL> grant create session,create table to u1;
Grant succeeded.

查询用户权限
SQL> select * from dba_sys_privs where grantee = 'U1';
GRANTEE                  PRIVILEGE                ADMIN_OPTION
------------------------ ------------------------ ------------
U1                       CREATE SESSION           NO
U1                       CREATE TABLE             NO
```

【例】 授予用户 u1 create view 权限，带 admin option 选项，通过 u1 将 create view 权限授予 u2。

```
C:\> sqlplus system/oracle@remote

授予 u1 权限
SQL> grant create view to u1 with admin option;
Grant succeeded.

查询用户权限
SQL> select * from dba_sys_privs where grantee in ('U1', 'U2');
GRANTEE                  PRIVILEGE                ADMIN_OPTION
------------------------ ------------------------ ------------
U1                       CREATE SESSION           NO
U1                       CREATE TABLE             NO
U1                       CREATE VIEW              YES
U2                       CREATE SESSION           NO

通过 u1 将 create view 权限授予 u2
C:\> sqlplus u1/u1@remote
SQL> grant create view to u2;

SQL> select * from dba_sys_privs where grantee in ('U1', 'U2');
GRANTEE                  PRIVILEGE                ADMIN_OPTION
------------------------ ------------------------ ------------
U1                       CREATE SESSION           NO
U1                       CREATE TABLE             NO
U1                       CREATE VIEW              YES
U2                       CREATE SESSION           NO
U2                       CREATE VIEW              NO
```

2. 权限回收

权限回收语法如下:

```
REVOKE { system_privilege [, …] | ALL PRIVILEGES } FROM grantee ;
```

属性说明:

1) system_privilege | ALL PRIVILEGES

指定要回收的系统权限,可同时回收多个系统权限,权限名用","隔开。也可以通过 ALL PRIVILEGES 将全部权限回收。

2) grantee

指定被授权者,可以是用户或角色。

5.2.2 对象权限

1. 权限授予

权限授予语法如下:

```
GRANT {object_privilege [, …] | ALL [ PRIVILEGES ] }
ON object
TO grantee
[ WITH GRANT OPTION ];
```

属性说明:

1) object_privilege | ALL [PRIVILEGES]

指定要授予用户或角色的对象权限,可同时授予多个对象权限,权限名用","隔开。也可以通过 ALL PRIVILEGES 将全部权限授予用户或角色。常用对象权限参见表 5-6。授权时不能与系统权限一起授予。

2) object

指定被授权的数据库对象,可以是表、视图等。

3) grantee

指定被授权者,可以是用户名或角色名,可同时授予多个用户或角色。

4) WITH GRANT OPTION

指定被授权者具有将权限授予其他用户或角色的权力,也可以从其他用户或角色回收该权限。

表 5-6 常用对象权限

对象类型	权限	描述
TABLE VIEW	DELETE	删除数据
	INSERT	插入数据
	SELECT	查询
	UPDATE	修改数据
SEQUENCE	SELECT	查询
PROCEDURE FUNCTION	EXECUTE	执行
DIRECTORY	READ	读
	WRITE	写

【例】 创建表 t01,创建用户 u1,将表 t01 的插入数据、查询数据权限授予 u1。

```
C:\> sqlplus system/oracle@remote

创建表
C:\> create table t01(id number);

创建用户
SQL> create user u1 identified by u1 quota unlimited on users;
User created.

授予权限
SQL> grant insert, select on t01 to u1;
Grant succeeded.

查询用户权限
SQL> select grantee, owner, table_name, grantor, privilege, grantable
  2   from dba_tab_privs where grantee = 'U1';
GRANTEE      OWNER      TABLE_NAME    GRANTOR      PRIVILEGE    GRANTABLE
-----------  ---------  ------------  -----------  -----------  ----------
U1           SYSTEM     T01           SYSTEM       SELECT       NO
U1           SYSTEM     T01           SYSTEM       INSERT       NO
```

【例】 授予 u1 用户 update 表 t01 权限,带 grant option 选项,通过 u1 用户将 update 表 t01 权限授予 u2 用户。

```
C:\> sqlplus system/oracle@remote

授予 u1 权限
SQL> grant update on t01 to u1 with grant option;
Grant succeeded.

查询用户权限
SQL> select grantee, owner, table_name, grantor, privilege, grantable
  2   from dba_tab_privs where grantee = 'U1';
GRANTEE      OWNER      TABLE_NAME    GRANTOR      PRIVILEGE    GRANTABLE
-----------  ---------  ------------  -----------  -----------  ----------
U1           SYSTEM     T01           SYSTEM       SELECT       NO
U1           SYSTEM     T01           SYSTEM       INSERT       NO
U1           YSTEM      T01           SYSTEM       UPDATE       YES

通过 u1 将 update 表 t01 权限授予 u2
C:\> sqlplus u1/u1@remote
SQL> grant update on system.t01 to u2;

SQL> select grantee, owner, table_name, grantor, privilege, grantable
  2   from dba_tab_privs where grantee in ('U1', 'U2');
GRANTEE      OWNER      TABLE_NAME    GRANTOR      PRIVILEGE    GRANTABLE
-----------  ---------  ------------  -----------  -----------  ----------
U1           SYSTEM     T01           SYSTEM       SELECT       NO
U1           SYSTEM     T01           SYSTEM       INSERT       NO
U1           SYSTEM     T01           SYSTEM       UPDATE       YES
U1           SYSTEM     T01           U1           UPDATE       NO
```

2. 权限回收

权限回收语法如下：

```
REVOKE { object_privilege [,…] | ALL [PRIVILEGES] }
ON object
FROM grantee ;
```

属性说明：

1) object_privilege | ALL [PRIVILEGES]

指定要回收的对象权限，可同时回收多个对象权限，权限用","隔开。也可以通过 ALL PRIVILEGES 将全部权限回收。但是不能同时回收多个对象的权限，而且也不能与系统权限一起回收，如下语句都不能执行：

```
revoke select on t01,t02 from u1;
revoke select on t01, create table from u1;
```

2) object

指定被授权的数据库对象，可以是表、视图等。

3) grantee

指定被授权者，可以是用户或角色。

5.2.3 相关视图

1. SYSTEM_PRIVILEGE_MAP

SYSTEM_PRIVILEGE_MAP 视图显示数据库中所有系统权限，参见表 5-7。

表 5-7 SYSTEM_PRIVILEGE_MAP 列介绍

列 名	描 述
PRIVILEGE	编码
NAME	权限名

2. DBA_SYS_PRIVS

DBA_SYS_PRIVS 视图显示数据库中用户或角色被授予的系统权限，参见表 5-8。

表 5-8 DBA_SYS_PRIVS 列介绍

列 名	描 述
GRANTEE	被授权者，用户名或角色名
PRIVILEGE	权限
ADMIN_OPTION	是否有管理权限

3. TABLE_PRIVILEGE_MAP

TABLE_PRIVILEGE_MAP 视图显示数据库中所有对象权限，参见表 5-9。

表 5-9 TABLE_PRIVILEGE_MAP 列介绍

列 名	描 述
PRIVILEGE	编码
NAME	权限名

4. DBA_TAB_PRIVS

DBA_TAB_PRIVS 视图显示数据库中用户或角色被授予的对象权限,参见表 5-10。

表 5-10 DBA_TAB_PRIVS 列介绍

列 名	描 述
GRANTEE	被授权者,用户名或角色名
OWNER	对象拥有者
TABLE_NAME	对象名
GRANTOR	授权者
PRIVILEGE	权限
GRANTABLE	是否有授予权限

5.3 角色管理

角色是一组权限的集合,可以是系统权限或对象权限。角色可以授予任何用户或角色,通过角色可以简化用户权限的管理,角色可以是 Oracle 数据库自带的,也可以自定义。

5.3.1 自定义角色

1. 创建角色

创建角色语法如下:

```
CREATE ROLE role
IDENTIFIED BY password;
```

属性说明:

1) role

指定要创建的角色名称。

2) IDENTIFIED BY

指定角色口令,当角色不是用户默认角色时,如通过 SET ROLE 命令激活角色时需要这个口令。

角色本身的权限授予与回收与用户权限的授予与回收一样,参见 5.2 节。

【例】 创建一个角色,并授予 CREATE SESSION,CREATE TABLE 系统权限以及查询 system 用户 t01 表权限。

```
C:\> sqlplus system/oracle@remote

创建角色
C:\> create role role1;
授予权限
SQL> grant create session, create table to role1;
Grant succeeded.

SQL> grant select on t01 to role1;
Grant succeeded.

SQL> select * from dba_sys_privs where grantee = 'ROLE1';
GRANTEE                PRIVILEGE            ADMIN_OPTION
--------------------   ------------------   -------------
ROLE1                  CREATE SESSION       NO
ROLE1                  CREATE TABLE         NO

SQL> select grantee,owner,table_name,grantor,privilege,grantable
    from dba_tab_privs where grantee = 'ROLE1';
GRANTEE   OWNER    TABLE_NAME   GRANTOR   PRIVILEGE   GRANTABLE
-------   ------   ----------   -------   ---------   ---------
ROLE1     SYSTEM   T01          SYSTEM    SELECT      NO
```

2. 删除角色

语法：

```
DROP ROLE role;
```

5.3.2 角色授予与回收

1. 角色授予

角色授予语法如下：

```
GRANT role [, … ] TO grantee [ WITH ADMIN OPTION ];
```

属性说明：

1) role

指定要授予的角色,可同时授予多个角色,角色用","隔开。常用系统角色参见表 5-11。

2) grantee

指定被授权者,可以是用户名或角色名,可同时授予多个用户或角色。

3) WITH ADMIN OPTION

指定被授权者具有管理该角色的权利,被授权者可以将该角色授予其他用户或角色,也可以从其他用户或角色回收该角色。

表 5-11 常用角色

列 名	描 述
CONNECT	只包含 connect session 权限
RESOURCE	包含 create table、create sequence 等基本系统权限
DBA	包含所有系统权限
SELECT_CATALOG_ROLE	包含查询数据字典权限
DELETE_CATALOG_ROLE	删除系统审计表 AUD$
EXECUTE_CATALOG_ROLE	执行数据库字典对象权限

2. 角色回收

角色回收语法如下：

```
REVOKE role [,…] FROM grantee ;
```

属性说明：

role 指定要回收的角色，可同时回收多个角色，角色间用","隔开。

5.3.3 相关视图

1. DBA_ROLES

DBA_ROLES 视图显示数据库中所有角色信息，参见表 5-12。

表 5-12 DBA_ROLES 列介绍

列 名	描 述
ROLE	角色名称
AUTHENTICATION_TYPE	授权机制

2. DBA_ROLE_PRIVS

DBA_ROLE_PRIVS 视图显示数据库所有角色的授予信息，参见表 5-13。

表 5-13 DBA_ROLE_PRIVS 列介绍

列 名	描 述
GRANTEE	被授予者
GRANTED_ROLE	授予的角色
ADMIN_OPTION	是否有管理权限
DEFAULT_ROLE	是否默认角色

3. ROLE_SYS_PRIVS

ROLE_SYS_PRIVS 视图显示数据库中角色被授予的系统权限，参见表 5-14。

表 5-14 ROLE_SYS_PRIVS 列介绍

列 名	描 述
ROLE	被授权角色
PRIVILEGE	权限
ADMIN_OPTION	是否有管理权限

4. ROLE_TAB_PRIVS

ROLE_TAB_PRIVS 视图显示数据库中角色被授予的对象权限,参见表 5-15。

表 5-15 ROLE_TAB_PRIVS 列介绍

列 名	描 述
ROLE	被授权角色名
OWNER	对象拥有者
TABLE_NAME	对象名
PRIVILEGE	权限
GRANTABLE	是否有授予权限

5. ROLE_ROLE_PRIVS

ROLE_ROLE_PRIVS 视图显示数据库所有角色的被授予信息,参见表 5-16。

表 5-16 ROLE_ROLE_PRIVS 列介绍

列 名	描 述
GRANTEE	被授予者
GRANTED_ROLE	授予的角色
ADMIN_OPTION	是否有管理权限

5.4 安全审计

审计用来监控并记录数据库用户的行为,是保障数据库安全的一种手段。审计信息可以记录在数据字典表中或操作系统文件中。审计可以分为标准审计和精度审计,标准审计可用来审计 SQL 语句、权限、Schema 对象以及网络活动,通过 AUDIT 语句进行设置,精度审计使用 Oracle 包来创建审计策略,可根据条件审计对象上的操作。

5.4.1 标准审计

1. 审计语句

审计哪些用户执行了某些 SQL 语句。

1) 增加审计语法

```
AUDIT { ALL | ALL STATEMENTS | sql_statement_shortcut }
[ BY user [, …] ]
[ BY ACCESS | SESSION ]
[ WHENEVER [NOT] SUCCESSFUL ];
```

属性说明:

(1) ALL | ALL STATEMENTS | sql_statement_shortcut

指定要审计的 SQL 语句,ALL 代表所有语句,ALL STATEMENTS 代表用户执行过的语句,sql_statement_shortcut 指定具体语句,比如 create table 或 table,如果是 table,表示审计所有语句中带 table 关键字的语句,比如 drop table、alter table 以及 create table。

(2) BY user

指定审计哪个或哪些用户的操作，多个用户名用","隔开。

(3) BY ACCESS | SESSION

BY ACCESS 指定每个操作都审计，单独记录一条日志，BY SESSION 指定每个会话针对一个 SQL 记录一条日志。

(4) WHENEVER [NOT] SUCCESSFUL

指定审计执行成功的语句或是不成功的语句，默认都审计。

【例】 审计用户 u1 执行创建表语句。

```
设置参数
C:\> sqlplus system/oracle@remote
SQL> alter system set audit_trail = db,extended scope = spfile;
SQL> shutdown immediate
SQL> startup

增加审计项
C:\> sqlplus system/oracle@remote
SQL> audit create table by u1;
Audit succeeded.

查询审计项
SQL> select user_name,audit_option,success,failure
    from dba_stmt_audit_opts;
USER_NAME      AUDIT_OPTION      SUCCESS
----------     ---------------   ----------   ---------------
U1             CREATE TABLE      BY ACCESS

执行操作
C:\> sqlplus u1/u1@remote
SQL> create table tbl01(id number);
Table created.

SQL> create table tbl01(id number);
create table tbl01(id number)
             *
ERROR at line 1:
ORA - 00955: name is already used by an existing object

查询审计结果
SQL> select owner,obj_name,action_name,returncode,sql_text
    from dba_audit_trail where username = 'U1';
OWNER  OBJ_NAME   ACTION_NAME      RETURNCODE   SQL_TEXT
-----  --------   -------------    ----------   ------------------------------
U1     TBL01      CREATE TABLE                  0 create table tbl01(id number)
U1     TBL01      CREATE TABLE                  955 create table tbl01(id number)
```

2）去除审计语法

```
NOAUDIT { ALL | ALL STATEMENTS | sql_statement_shortcut }
[ BY user [, … ] ]
[ WHENEVER [NOT] SUCCESSFUL ];
```

【例】 去除用户 u1 执行创建表语句审计项。

```
C:\> sqlplus system/oracle@remote
SQL> noaudit create table by u1;
Noaudit succeeded.

查询审计项
SQL> select user_name,audit_option,success,failure
     from dba_stmt_audit_opts;
no rows selected
```

2. 审计权限

审计哪些用户执行了某些权限。

1）增加审计语法

```
AUDIT { ALL PRIVILEGES | system_privileges }
[ BY user [, … ] ]
[ BY ACCESS | SESSION ]
[ WHENEVER [NOT] SUCCESSFUL ];
```

属性说明：

（1）ALL PRIVILEGES | system_privileges

指定要审计的权限，ALL PRIVILEGES 代表所有系统权限，system_privileges 指定具体权限，比如 select any table。

（2）BY user

指定审计哪个或哪些用户的操作，多个用户名用","隔开。

（3）BY ACCESS | SESSION

BY ACCESS 指定每个操作都审计，单独记录一条日志，BY SESSION 指定每个会话针对一个 SQL 记录一条日志。

（4）WHENEVER [NOT] SUCCESSFUL

指定审计执行成功的语句或是不成功的语句，默认都审计。

【例】 审计用户 u1 访问 system 用户的 t01 表。

```
授权、增加审计项
C:\> sqlplus system/oracle@remote
SQL> grant select any table to u1;
SQL> audit select any table by u1;

查询审计项
SQL> select user_name,privilege,success,failure from dba_priv_audit_opts;
```

```
USER_NAME         PRIVELEGE                    SUCCESS
---------------   -------------------------    ----------   -----------
U1                SELECT ANY TABLE             BY SESSION BY SESSION
```

执行操作
```
C:\> sqlplus u1/u1@remote
SQL> select * from system.t01;
no rows selected
```

查询审计结果
```
SQL> select owner,obj_name,action_name,returncode,sql_text
    from dba_audit_trail where username = 'U1';
OWNER     OBJ_NAME      ACTION_NAME    RETURNCODE    SQL_TEXT
------    ---------     -----------    ----------    ------------------------
SYSTEM    T01           SESSION REC                  0 select * from system.t01
```

2）去除审计语法

```
NOAUDIT { ALL PRIVILEGES | system_privileges }
[ BY user [, …] ]
[ WHENEVER [NOT] SUCCESSFUL ];
```

【例】 去除用户 u1 执行创建表语句审计项。

去除审计项
```
C:\> sqlplus system/oracle@remote
SQL> noaudit select any table by u1;
Noaudit succeeded.
```

查询审计项
```
SQL> select user_name,audit_option,success,failure
    from dba_priv_audit_opts;
no rows selected
```

3. 审计对象

审计哪些用户在哪些对象上执行了操作。

1）增加审计语法

```
AUDIT { ALL | sql_operation }
ON object
[ BY ACCESS | SESSION ]
[ WHENEVER [NOT] SUCCESSFUL ];
```

属性说明：

（1）ALL | sql_operation

指定要审计对象上的 SQL 操作，ALL 代表所有操作语句，sql_operation 指定具体语句，比如 insert、delete 等。

(2) BY ACCESS | SESSION

BY ACCESS 指定每个操作都审计,单独记录一条日志,BY SESSION 指定每个会话针对一个 SQL 记录一条日志。

(3) WHENEVER [NOT] SUCCESSFUL

指定审计执行成功的语句或是不成功的语句,默认都审计。

【例】 审计用户 u1 执行创建表语句。

```
增加审计项
C:\> sqlplus system/oracle@remote
SQL> grant insert on system.t01 to u1;
SQL> audit insert on system.t01;
Audit succeeded.

查询审计项
SQL> select owner,object_name,del,ins,sel,upd from dba_obj_audit_opts;
OWNER    OBJECT_NAME    DEL    INS    SEL    UPD
------   -----------    ----   ----   ----   ----
SYSTEM   T01            -/-    S/S    -/-    -/-

执行操作
C:\> sqlplus u1/u1@remote
SQL> insert into system.t01 values(100);
1 row created.

查询审计结果
SQL> select owner, obj_name, returncode, sql_text
     from dba_audit_trail where username = 'U1';
OWNER    OBJ_NAME    RETURNCODE    SQL_TEXT
------   --------    ----------    --------
SYSTEM   T01         0             insert into system.t01 values(100)
```

2) 去除审计语法

```
NOAUDIT { ALL | sql_operation }
ON object
[ BY ACCESS | SESSION ]
[ WHENEVER [NOT] SUCCESSFUL ];
```

【例】 去除用户 u1 执行创建表语句审计项。

```
去除审计项
C:\> sqlplus system/oracle@remote
SQL> noaudit insert on system.t01;
Noaudit succeeded.

查询审计项
SQL> select owner,object_name,del,ins,sel,upd from dba_obj_audit_opts;
no rows selected
```

5.4.2 精度审计

精度审计是指根据条件对数据库对象进行审计,不审计所有操作,审计结果更有针对性。精度审计通过 Oracle 包 DBMS_FGA 中过程实现,与系统参数 AUDIT_TRAIL 无关。

1) 增加审计策略

```
DBMS_FGA.ADD_POLICY(
    object_schema        VARCHAR2,
    object_name          VARCHAR2,
    policy_name          VARCHAR2,
    audit_condition      VARCHAR2,
    audit_column         VARCHAR2,
    enable               BOOLEAN,
    statement_types      VARCHAR2
);
```

参数说明见表 5-17。

表 5-17 DBMS_FGA.ADD_POLICY 参数说明

参数	描述
object_schema	审计对象的 Schema
object_name	审计的对象名
policy_name	审计策略名
audit_condition	审计条件
audit_column	审计列
enable	启用
statement_types	审计语句

【例】 审计用户 u1 中工资表工资大于 5000 的操作,观察审计结果。

```
增加审计项
C:\> sqlplus u1/u1@remote
SQL> create table salary(id number,name varchar2(20),sal number);
Table created.

SQL> begin
        dbms_fga.add_policy(
            object_schema    => 'u1',
            object_name      => 'salary',
            policy_name      => 'p1',
            audit_condition  => 'sal>5000',
            statement_types  => 'select,insert,delete,update'
        );
    end;
    /
PL/SQL procedure successfully completed.
```

```
执行数据操作
SQL > insert into salary values(1,'a',3000);
1 row created.

SQL > insert into salary values(2,'b',8000);
1 row created.

查询审计项
SQL > select db_user, sql_text from dba_fga_audit_trail;
DB_USER         SQL_TEXT
-----------     --------------------------------------------
U1              insert into salary values(2,'b',8000)
```

2) 删除审计策略

```
DBMS_FGA.DROP_POLICY(
    object_schema    VARCHAR2,
    object_name      VARCHAR2,
    policy_name      VARCHAR2
);
```

【例】 删除审计策略。

```
删除审计策略
C:\> sqlplus u1/u1@remote
SQL > begin
        dbms_fga.drop_policy(
            object_schema => 'u1',
            object_name => 'salary',
            policy_name => 'p1'
        );
      end;
      /
PL/SQL procedure successfully completed.
```

5.4.3 相关视图

1. DBA_STMT_AUDIT_OPTS

DBA_STMT_AUDIT_OPTS 视图显示语句审计信息，参见表 5-18。

表 5-18 DBA_STMT_AUDIT_OPTS 列介绍

列 名	描 述
USER_NAME	被审计用户名
AUDIT_OPTION	审计项
SUCCESS	是否审计执行成功
FAILURE	是否审计执行失败

2. DBA_PRIV_AUDIT_OPTS

DBA_PRIV_AUDIT_OPTS 视图显示权限审计信息，参见表 5-19。

表 5-19　DBA_PRIV_AUDIT_OPTS 列介绍

列　名	描　述
USER_NAME	被审计用户名
PRIVILEGE	审计项
SUCCESS	是否审计执行成功
FAILURE	是否审计执行失败

3. DBA_OBJ_AUDIT_OPTS

DBA_OBJ_AUDIT_OPTS 视图显示对象审计信息，参见表 5-20。

表 5-20　DBA_OBJ_AUDIT_OPTS 列介绍

列　名	描　述
OWNER	对象所属用户
OBJECT_NAME	对象名
DEL	是否审计 DELETE 语句
INS	是否审计 INSERT 语句
SEL	是否审计 SELECT 语句
UPD	是否审计 UPDATE 语句

4. DBA_AUDIT_TRAIL

DBA_AUDIT_TRAIL 视图显示审计结果信息，参见表 5-21。

表 5-21　DBA_AUDIT_TRAIL 列介绍

列　名	描　述
USERNAME	用户名
OBJ_NAME	对象名
ACTION_NAME	动作名
RETURNCODE	执行结果代码
EXTENDED_TIMESTAMP	审计时间
SQL_BIND	绑定变量
SQL_TEXT	SQL 文本

5. DBA_AUDIT_POLICY

DBA_AUDIT_POLICY 视图显示审计策略信息，参见表 5-22。

表 5-22　DBA_AUDIT_POLICY 列介绍

列　名	描　述
OBJECT_SCHEMA	对象所属用户 Schema
OBJECT_NAME	对象名
POLICY_NAME	策略名
POLICY_TEXT	审计条件
DEL	是否审计 DELETE 语句
INS	是否审计 INSERT 语句
SEL	是否审计 SELECT 语句
UPD	是否审计 UPDATE 语句

6. DBA_FGA_AUDIT_TRAIL

DBA_FGA_AUDIT_TRAIL 视图显示精度审计结果信息,参见表 5-23。

表 5-23　DBA_FGA_AUDIT_TRAIL 列介绍

列　名	描　述
DB_USER	指定 SQL 用户
OBJECT_SCHEMA	对象所属用户
OBJECT_NAME	对象名
POLICY_NAME	策略名
SQL_TEXT	SQL 文本
SQL_BIND	绑定变量
EXTENDED_TIMESTAMP	审计时间

小　　结

本章讲解了 Oracle 数据库的用户权限管理,包括 4 部分内容:一是用户管理,包括用户信息的维护(创建、修改和删除),概要文件的作用和使用;二是权限管理,分别介绍了常用的系统权限和对象权限,权限的授予和回收;三是角色管理,包括角色的创建、角色的权限授予和回收;四是安全审计,包括标准审计和精度审计,标准审计中讲解了权限审计、语句审计和对象审计。

思　考　题

1. Oracle 数据库用户验证有几种方式?
2. 创建一个用户 u1,密码 abc,默认表空间 tbs,写出创建语句。
3. 授予用户 u1 创建表、创建视图权限,写出授权语句。
4. 创建概要文件 myprof,增加限制项 SESSIONS_PER_USER 和 FAILED_LOGIN_ATTEMPTS,然后将用户 u1 的概要文件指定为 myprof。
5. 创建密码验证函数 myfunc,自定义验证规则,然后修改 myprof 增加验证函数,之后修改 u1 用户密码进行测试。
6. 创建角色 r1,授予角色 r1 创建表和创建视图权限,回收用户 u1 的系统权限,将角色 r1 授予 u1,写出所有操作语句。
7. 审计用户 u1 对表的操作,包括创建表、删除表,写出所有相关操作步骤和语句。
8. 假设存在表 salary(id number,name varchar2(20),salary number),对工资大于 5000 的数据变化进行审计,写出审计语句。

第 6 章　REDO 与 UNDO

Oracle 数据库记录的事务日志称为 REDO，事务数据的前映像称为 UNDO。本章将重点讲解 REDO 和 UNDO 在 Oracle 数据库中的重要地位、基本原理以及维护命令。

6.1　REDO

6.1.1　REDO 概述

Oracle 数据库中包含大量数据文件，里面存储着关键的业务数据，数据文件的损坏将导致数据丢失，这就需要一种数据结构来保护数据文件的安全，一旦数据文件损坏可以恢复，这种结构就是 REDO 日志，也就是重做日志。Oracle 数据库中的任何数据变化都是先记录日志，在检查点发生时再更新数据文件，日志文件是顺序写，而数据文件是离散写，日志文件写效率高于数据文件写。

Oracle 数据库中 REDO 相关组件包括 SGA 中的日志缓冲区 LOG BUFFER、后台进程 LGWR 以及在线日志文件，如图 6-1 所示。

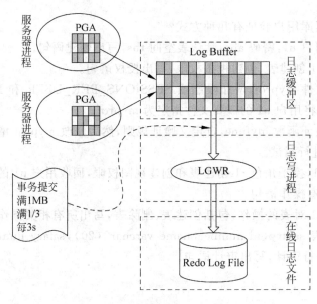

图 6-1　日志组件关系图

日志缓冲区是 SGA 中的一个循环使用区域,存储描述数据库变化的 REDO 条目,REDO 条目包含必要的信息重新构建或重做一遍由 DML 或 DDL 语句造成的数据库变化,数据库恢复在数据文件上应用 REDO 信息构建丢失的数据。

写 REDO 信息的过程主要分成以下两个步骤。

1. 服务器进程复制日志记录

记录数据库变化的 REDO 条目首先在服务器进程的 PGA 中缓存,然后将 REDO 条目复制到 SGA 中的日志缓冲区。日志的块大小与操作系统块大小一致,默认 512B。

2. 后台进程 LGWR 写日志文件

当条件满足时,后台进程 LGWR 将日志缓冲区中内容写出到当前状态(current)的在线日志文件。

触发条件如下。

(1) 用户提交事务。

(2) 日志切换。

(3) 每 3s。

(4) 日志缓冲区缓冲日志记录达到 1/3 满或满 1MB 日志数据。

(5) 在 DBWn 写出修改数据前。

6.1.2 日志组

Oracle 数据库使用多组日志来记录数据库变化,日志组本身大小固定,数据库循环使用每个日志组,数据库同一时间只写一个日志组的文件,当写满后,自动切换到下一组。数据库创建时默认有三组日志,如图 6-2 所示。

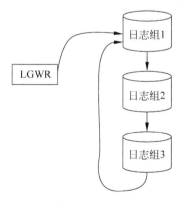

图 6-2 日志组

鉴于日志信息的重要性,每个日志组可设置多个日志成员,Oracle 数据库写日志时同时写日志组内所有成员文件,保持文件内容一致,每组中只要有一个文件可用,数据库将继续工作。通常将多个成员进行物理分散,防止同时损坏,如图 6-3 所示。

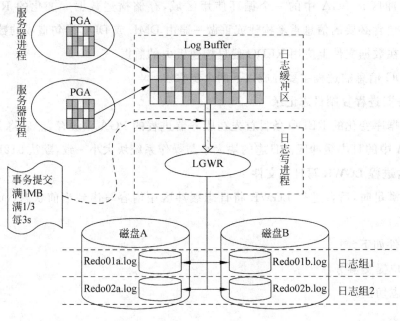

图 6-3 多个成员日志组

6.1.3 日志维护

1. 设置归档

归档是指日志组切换后,在日志文件被覆写前由后台进程 ARCn 将日志文件内容复制到一个归档文件,对日志内容进行保护。例如,日志组由组 A 切换到组 B,在组 A 文件被覆写前,ARCn 将组 A 的内容进行复制。数据库可以运行在归档模式或非归档模式,可在数据库 MOUNT 状态下改变归档模式。

设置数据库归档模式语法如下:

```
ALTER DATABASE { ARCHIVELOG | NOARCHIVELOG }
```

1) ARCHIVELOG

将数据库设置为归档模式。

2) NOARCHIVELOG

将数据库设置为非归档模式。

【例】 将数据库设置为归档模式。

```
C:\> sqlplus sys/oracle@remote as sysdba
查看归档状态
SQL > archive log list;
Database log mode              No Archive Mode
Automatic archival             Disabled
Archive destination            USE_DB_RECOVERY_FILE_DEST
Oldest online log sequence     152
Current log sequence           154
```

```
SQL > select log_mode from v $ database;
LOG_MODE
------------
NOARCHIVELOG

设置归档
SQL > shutdown immediate
Database closed.
Database dismounted.
ORACLE instance shut down.

SQL > startup mount
ORACLE instance started.
Total System Global Area   368263168 bytes
Fixed Size                   1374668 bytes
Variable Size              314574388 bytes
Database Buffers            46137344 bytes
Redo Buffers                 6176768 bytes
Database mounted.

SQL > alter database archivelog;
Database altered.

SQL > select log_mode from v $ database;
LOG_MODE
------------
ARCHIVELOG

SQL > alter database open;
Database altered.
```

2. 设置强制日志模式

将数据库设置为强制日志模式,则数据库中的任何变化都要记录日志,临时表空间中的数据变化不记录日志。

设置强制日志模式语法如下:

```
ALTER DATABASE { FORCE LOGGING | NO FORCE LOGGING }
```

1) FORCE LOGGING

将数据库设置为强制日志模式。

2) NO FORCE LOGGING

将数据库设置为非强制日志模式。

【例】 将数据库设置为强制日志模式。

```
C:\> sqlplus sys/oracle@remote as sysdba
查看强制日志状态
SQL> select force_logging from v$database;
FORCE_LOGGING
---------------------------
NO

设置强制日志模式
SQL> alter database force logging;
Database altered.

SQL> select force_logging from v$database;
FORCE_LOGGING
---------------------------
YES
```

3. 增删日志组

数据库创建时默认包含三组日志,可根据实际情况对日志组进行添加或删除,数据库中至少要包含两组日志。

增删日志组语法如下:

```
ALTER DATABASE
{ ADD LOGFILE  [GROUP integer]  (filename[,…])  SIZE size
 |  DROP LOGFILE GROUP integer}
```

1) ADD LOGFILE

增加日志组时可指定组号,如果未指定则系统自动分配,每组可包含多个成员,多个成员间用","隔开,最后通过 SIZE 语句指定日志组所有成员文件大小,同一组的成员大小相同。

2) DROP LOGFILE GROUP

通过日志组编号删除日志组。

【例】 添加日志组。

```
C:\> sqlplus sys/oracle@remote as sysdba
查看日志组
SQL> select group# from v$log;
GROUP#
----------
     1
     2
     3
增加日志组
SQL> alter database add logfile group 4 ('redo04a.log','redo04b.log') size 20M;
Database altered.

SQL> select group# from v$log;
GROUP#
```

```
----------
         1
         2
         3
         4
```
删除日志组
```
SQL> alter database drop logfile group 4;
Database altered.
```

4. 增删日志成员

数据库创建时默认每组包含一个成员,可增加成员,数据库保持同组内成员文件内容一致,只要每组有一个成员文件可用,数据库即可正常使用。

增删日志成员语法如下:

```
ALTER DATABASE
{ ADD LOGFILE MEMBER 'filename' TO GROUP integer
  | DROP LOGFILE MEMBER 'filename'}
```

1) ADD LOGFILE MEMBER

向指定的日志组添加成员。

2) DROP LOGFILE MEMBER

删除指定的成员文件。

【例】 添加日志成员。

```
C:\> sqlplus sys/oracle@remote as sysdba
查看组成员
SQL> select group#,member from v$logfile;
GROUP#     MEMBER
-------    -----------------------------------
1          REDO01a.LOG
2          REDO02a.LOG
3          REDO03a.LOG
添加成员
SQL> alter database add logfile member 'redo03b.log' to group 3;
Database altered.

SQL> select group#,member from v$logfile;
GROUP#     MEMBER
-------    -----------------------------------
1          REDO01a.LOG
2          REDO02a.LOG
3          REDO03a.LOG
3          REDO03b.LOG
删除成员
SQL> alter database drop logfile member 'redo03b.log';
Database altered.
```

6.1.4 相关视图

1. V$LOG

V$LOG 视图显示数据库中日志文件组信息,参见表 6-1。

表 6-1 V$LOG 列介绍

列 名	描 述
GROUP#	日志组号
SEQUENCE#	日志序列号
BYTES	日志大小
MEMBERS	成员个数
ARCHIVED	是否归档
STATUS	日志文件状态
FIRST_CHANGE#	记录数据库变化对应的最低 SCN 号

2. V$LOGFILE

V$LOGFILE 视图显示数据库日志文件信息,参见表 6-2。

表 6-2 V$LOGFILE 列介绍

列 名	描 述
GROUP#	日志组号
STATUS	文件状态
MEMBER	文件名

3. V$LOG_HISTORY

V$LOG_HISTORY 视图显示数据库日志文件历史信息,参见表 6-3。

表 6-3 V$LOG_HISTORY 列介绍

列 名	描 述
SEQUENCE#	日志序列号
FIRST_CHANGE#	数据库变化起始 SCN 号
NEXT_CHANGE#	数据库变化结束 SCN 号

6.2 UNDO

6.2.1 UNDO 概述

Oracle 数据库在数据发生变化时不但要记录 REDO 信息,同时要记录 UNDO 信息,REDO 信息用来进行数据恢复,重新产生数据变化,而 UNDO 信息用来撤销数据的变化。比如:向表 T01 中插入一条数据,记录的 REDO 信息是 INTERT INTO T01 VALUES (…),而记录的 UNDO 信息是 DELETE FROM T01 WHERE ROWID='…'。

Oracle 数据库使用回滚段来记录 UNDO 信息,回滚段是数据库对象,是用来保持数据变化前数据从而提供一致读和保障事务完整性的一段磁盘存储区域。Oracle 9i 以前回滚段是手动管理,需要数据库管理员进行维护,回滚段的数量及大小都需要进行认真计划并在运行环境中根据实际情况不断调整。Oracle 9i 以后版本使用 UNDO 表空间由数据库自动进行管理,回滚段由原来的名称改为 UNDO 段。当一个事务开始的时候,首先要给事务分配 UNDO 段,然后把变化前的数据和变化后的数据先写入日志缓冲区,并把变化前的数据写入 UNDO 段(这部分数据称为前映像),最后才在数据缓冲区中修改,如图 6-4 所示。

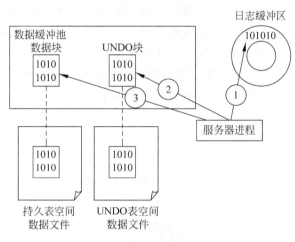

图 6-4 事务数据修改

关于回滚段的数据,针对不同操作记录的数据量是不同的,对于 INSERT 操作,对应的 UNDO 数据是根据 ROWID 进行 DELETE 操作,对于 UPDATE 操作,对应的 UNDO 数据是字段修改前的数据,对于 DELETE 操作,对应的 UNDO 数据是该行所有字段的原数据。在这三种操作中,DELETE 操作需要存储 UNDO 数据最多。

Oracle 数据库中 UNDO 数据有如下 4 个作用。

1. 事务回滚

当事务开始后,所有数据修改前都将前映像数据存储在 UNDO 段中,当用户执行 ROLLBACK 命令时,Oracle 数据库自动应用与事务关联的 UNDO 数据进行修改撤销,还原成数据修改前的状态。

2. 一致性查询

当会话 A 查询数据且未结束前,会话 B 修改了会话 A 尚未查询到的数据,这时会话 A 需要查询到会话 B 修改前的数据,Oracle 数据库将利用 UNDO 信息构建一个一致性查询数据块供会话 A 使用。

3. 实例恢复

Oracle 数据库在异常宕机情况下(比如突然掉电),数据缓冲池中的数据尚未来得及写到数据文件,当数据库再次启动时,数据库需要借助 REDO 信息进行数据库恢复。REDO 信息中包含提交数据以及未提交数据,实例恢复过程中数据库应用所有 REDO 信息,这时数据库中包含已经提交的数据以及未提交的数据,然后再应用 UNDO 信息将未提交的数据进行回滚。

4. 数据闪回

数据闪回是指将数据恢复到过去某个时间点,这时数据库将借助 UNDO 信息将数据回退到过去。

6.2.2 UNDO 段空间使用

Oracle 数据库利用 UNDO 段存储 UNDO 数据,UNDO 段在 UNDO 表空间中进行空间分配。事务开始时,数据库给事务分配 UNDO 段,在 UNDO 段的段头记录每个事务存储 UNDO 数据的数据块。多个事务可并发使用同一个 UNDO 段,可使用 UNDO 段的同一个区,但一个 UNDO 段的数据块只包含一个事务的数据。

UNDO 段在使用过程中,段中的区循环使用,当前区写满后,将使用下一个区。图 6-5 显示事务 T1 将 UNDO 段第三个区写满后写向第四个区。

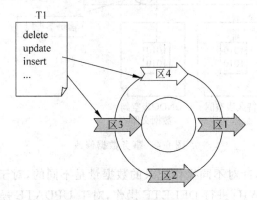

图 6-5　区循环使用

如果当前区写满后,下个区包含未提交的事务,则下个区不能被覆写,这时 Oracle 将分配一个新区,UNDO 段的区之间像个单项链表,分配新区后,将链接到当前区和下个区之间,如图 6-6 所示。

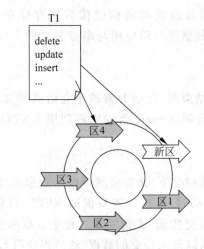

图 6-6　新区分配

6.2.3 UNDO 管理

从 Oracle 9i 开始,推荐使用 UNDO 表空间,由系统自动管理 UNDO 段,与 UNDO 相关的几个参数如下。

```
C:\> sqlplus sys/oracle@remote as sysdba
sql> show parameter undo
undo_management                      string       AUTO
undo_retention                       integer      900
undo_tablespace                      string       UNDOTBS1
```

1. UNDO_TABLESPACE

用于指定实例所要使用的 UNDO 表空间。设置的时候,必须保证该表空间存在,否则会导致实例启动失败。

使用 RAC 的时候,因为一个 UNDO 不能由多个实例同时使用,所以必须为每个实例配置一个独立的 UNDO 表空间。

2. UNDO_RETENTION

指定事务结束后 UNDO 信息在 UNDO 表空间中默认保留时间,可以通过设置 UNDO 表空间的强制保留(RETENTION GUARANTEE)防止 UNDO 信息在指定时间内被覆写。

保留时间与存储空间是紧密相关的,如果 UNDO 表空间的存储空间有限,那么 Oracle 回收已提交事务占用的空间,置 UNDO_RETENTION 于不顾。Oracle 9i 中如果出现空间不足,完成的事务信息就将被覆盖,这将影响一些一致性读和闪回功能。Oracle 10g 开始增加了 GUARANTEE 控制,可以指定 UNDO 表空间必须满足 UNDO_RETENTION 的限制。

3. UNDO_MANAGEMENT

用于指定 UNDO 数据的管理方式。如果使用自动管理,必须设置为 AUTO;如果手动管理,必须设置为 MANUAL。

如果使用自动管理模式,必须建立 UNDO 表空间,并且配置 UNDO_TABLESPACE 参数,否则 Oracle 会使用 SYSTEM 回滚段存放 UNDO 记录,并在警告日志记录中提示。

【例】 创建 UNDO 表空间并设置为系统 UNDO 表空间。

```
C:\> sqlplus sys/oracle@remote as sysdba

SQL> create undo tablespace undotbs datefile 'undotbs.dbf' size 100M
     autoextend on next 5M;
SQL> alter systm set undo_tablespace = undotbs;
System altered.
```

【例】 计算平均事务回滚率。

```
SQL > select value from v $ sysstat where name = 'user commits';
SQL > select value from v $ sysstat where name = 'user rollbacks';

计算公式：user rollbacks/(user commits + user rollbacks)
SQL > select s1.value commit,s2.value rollback,
(s2.value/(s1.value + s2.value)) * 100
from v $ sysstat s1,v $ sysstat s2
where s1.name = 'user commits' and s2.name = 'user rollbacks';
```

【例】 UNDO 表空间大小计算。

设置 UNDO 表空间的大小公式：
undo Size = Undo_retention * UPS

UPS 是 Undo Block per Second 的缩写，可以通过 v $ undostat.undoblks 获得，视图 v $ undostat 统计每 10min 生成 UNDO 块数。
```
SQL > select max(undoblks)/(10 * 60) UPS from v $ undostat;
```

根据上面查询得出的平均每秒生成 UNDO 块数乘以 undo_retention,再根据 UNDO 块大小得出表空间大小。
例如：平均每秒生成 100 块 UNDO 数据，假设数据库默认块大小为 8KB,Undo_retention 为 900s,那么 UNDO 表空间至少需要 900×100×8KB = 720MB。

6.2.4 相关视图

1. V $ ROLLNAME

V $ ROLLNAME 视图显示数据库中 UNDO 段名称，参见表 6-4。

表 6-4　V $ ROLLNAME 列介绍

列　　名	描　　述
USN	UNDO 段号
NAME	UNDO 段名

2. V $ ROLLSTAT

V $ ROLLSTAT 视图显示数据库 UNDO 段统计信息，参见表 6-5。

表 6-5　V $ ROLLSTAT 列介绍

列　　名	描　　述
USN	UNDO 段号
EXTENTS	UNDO 段包含的区数
RSSIZE	段大小
WRITES	写入字节数

续表

列　名	描　述
XACTS	活动事务数
GETS	读段头次数
WAITS	读段头等待次数
OPTSIZE	UNDO 段的最优大小
HWMSIZE	高水位线
SHRINKS	UNDO 段收缩次数
WRAPS	UNDO 段从头写次数
EXTENDS	UNDO 段扩展次数
AVESHRINK	平均收缩大小
AVEACTIVE	平均活动区大小
CUREXT	当前区
CURBLK	当前块

3. DBA_ROLLBACK_SEGS

DBA_ROLLBACK_SEGS 视图显示数据库 UNDO 段信息，参见表 6-6。

表 6-6　DBA_ROLLBACK_SEGS 列介绍

列　名	描　述
SEGMENT_NAME	UNDO 段名
OWNER	所有者
TABLESPACE_NAME	表空间名
FILE_ID	文件编号
BLOCK_ID	段头块

4. DBA_UNDO_EXTENTS

DBA_UNDO_EXTENTS 视图显示数据库 UNDO 段中区的信息，参见表 6-7。

表 6-7　DBA_UNDO_EXTENTS 列介绍

列　名	描　述
SEGMENT_NAME	UNDO 段名
OWNER	所有者
TABLESPACE_NAME	表空间名
EXTENT_ID	区号
FILE_ID	所在文件号
BLOCK_ID	起始块号
BLOCKS	块数
STATUS	区状态 ACTIVE：活动 EXPIRED：超过保留时间 UNEXPIRED：未超过保留时间

5. V$UNDOSTAT

V$UNDOSTAT 视图显示数据库 UNDO 段每 10min 的统计信息,参见表 6-8。

表 6-8　V$UNDOSTAT 列介绍

列　名	描　述
BEGIN_TIME	统计开始时间
END_TIME	统计结束时间
UNDOTSN	UNDO 表空间
UNDOBLKS	使用的 UNDO 块数
TXNCOUNT	发生事务数
MAXQUERYLEN	最长查询时间
MAXQUERYID	最长查询的 SQL ID
MAXCONCURRENCY	最大事务并发度
UNXPSTEALCNT	尝试获得未到期的次数
UNXPBLKRELCNT	释放未到期数据的块数
UNXPBLKREUCNT	重新使用未到期的块数
EXPSTEALCNT	尝试获得到期块的次数
EXPBLKRELCNT	释放到期数据的块数
EXPBLKREUCNT	重新使用到期的块数
SSOLDERRCNT	发生快照过旧错误的次数
NOSPACEERRCNT	发生空间不足错误的次数
ACTIVEBLKS	活动块数
UNEXPIREDBLKS	未到期块数
EXPIREDBLKS	到期块数

6.3　检 查 点

大多数关系型数据库都采用"在提交时并不强迫针对数据块的修改完成",而是"提交时保证修改记录(以重做日志的形式)写入日志文件"的机制,来获得性能的优势。这句话的另外一种描述是:当用户提交事务,写数据文件是"异步"的,写日志文件是"同步"的。

这就可能导致数据库实例崩溃时,内存中的数据缓冲区中修改过的数据,可能没有写入到数据文件中。数据库在重新打开时,需要进行实例恢复,确保已经提交的数据被写入到数据块中。检查点(CHECKPOINT)是这个过程中的重要机制,通过它来确定恢复时哪些重做日志应该被扫描并应用于恢复。

检查点是一个数据库事件,由后台进程 CKPT 触发。检查点发生时,CKPT 要做两件事:①通知 DBWn 进程将脏数据写出到数据文件上;②更新数据文件头及控制文件上的检查点信息。

检查点分为完全检查点、文件检查点以及增量检查点。

1. 完全检查点

指系统将数据缓冲区中所有数据文件的数据写出。

完全检查点触发条件:

1) shutdown 数据库（abort 方式除外）

关闭数据库时，触发完全检查点，数据缓冲区中的数据将全部写出到数据文件。

2) 数据库管理员手动请求

通过执行 ALTER SYSTEM CHECKPOINT 来触发。

3) 日志切换时

日志自动切换或执行 ALTER SYSTEM SWITCH LOGFILE 语句，会触发检查点。

2. 文件检查点

文件检查点指针对单个表空间，只将该表空间对应的数据文件在数据缓冲区中的数据写出。

文件检查点触发条件：

1) 将表空间离线

执行 ALTER TABLESPACE tablespace_name OFFLINE 语句，系统将表空间 tablespace_name 的所有数据文件内容写出。

2) 将表空间置成热备份状态

执行 ALTER TABLESPACE tablespace_name BEGIN BACKUP 语句，系统将表空间 tablespace_name 的所有数据文件内容写出。

3) 将表空间置成只读方式

执行 ALTER TABLESPACE tablespace_name READ ONLY 语句，系统将表空间 tablespace_name 的所有数据文件内容写出。

3. 增量检查点

增量检查点指检查点以增量方式完成，通过设置参数控制。

增量检查点设置方式：

Oracle 8i 前可通过 LOG_CHECKPOINT_INTERVAL 指定两次检查点间产生的 REDO 块最大数量，或通过 LOG_CHECKPOINT_TIMEOUT 指定两次检查点的时间间隔来控制检查点的发生。Oracle 9i 后通过参数 FAST_START_MTTR_TARGET 设置实例恢复时间，由系统来进行估算检查点发生时机，参数 FAST_START_MTTR_TARGET 设置时间单位为秒。

```
C:\> sqlplus sys/oracle@remote as sysdba

SQL> show parameter fast_start_mttr_target
NAME                                 TYPE        VALUE
------------------------------------ ----------- ------------------------------
fast_start_mttr_target               integer     0

SQL> alter system set fast_start_mttr_target = 30;
System altered.
```

小　结

本章主要讲解 REDO 和 UNDO，另外介绍了检查点。REDO 部分讲解了 REDO 日志作用、REDO 相关组件介绍、REDO 日志组和日志成员的维护；UNDO 部分讲解了 UNDO 作用，UNDO 段空间使用，以及 UNDO 表空间创建和相关参数介绍；检查点部分讲解了检查点作用，检查点的分类以及增量检查点设置方法。

思　考　题

1. LGWR 将日志缓冲区内容写出到日志文件的触发条件有哪些？
2. 新增一个日志组，包括两个成员，成员大小 100MB，写出创建语句。
3. UNDO 有哪些作用？
4. 创建一个 UNDO 表空间 undo1，并将该表空间设置为系统使用的 UNDO 表空间，写出相关语句。
5. 如何估算 UNDO 表空间大小？写出具体步骤。
6. 完全检查点有哪些触发条件？
7. 如何设置增量检查点？举例说明。

第 7 章　备份恢复与闪回技术

数据的安全可靠对任何一个企业都是至关重要的,数据库的备份恢复技术是保障数据安全的一种重要手段。本章重点介绍备份恢复原理、RMAN 工具使用以及闪回技术的运用。

7.1　备份恢复概述

数据库备份恢复可以手动进行,也可借助于工具。手动备份需要借助操作系统命令进行文件复制,可在数据库关闭状态下进行"冷备份",如果数据库处于归档模式,也可以在数据库打开状态下进行"热备份",手动恢复需要借助操作系统命令还原备份,再使用 SQL * Plus 命令完成恢复。手动备份效率不高,现在大多都采用 RMAN 工具进行备份恢复,本节主要介绍手动备份恢复。

7.1.1　备份文件

备份的目的是为了保障数据文件损坏或介质故障后数据库数据能够得到恢复,从而保障数据安全。Oracle 数据库包含 6 类文件,首先需要分析哪些文件需要备份,分析如下。

1. 数据文件

数据文件是 Oracle 数据库中存储数据的文件,比如表数据、索引数据等,是需要重点保护的对象。除了临时表空间外,持久表空间和 UNDO 表空间的数据文件都需要定期备份。

2. 在线日志文件

Oracle 使用在线日志文件组来记录数据库的变化,日志内容可以通过归档方式进行备份,在线日志文件通过冗余设置来保障安全,一般每组设置两个或多个成员,每组中只要有一个成员文件可用,数据库就可以正常工作。

3. 控制文件

控制文件记录数据库的状态以及结构信息,通过冗余设置来保障安全,一般设置两个控制文件,任意一个损坏数据库将不能工作,需要定期备份,当数据库结构发生变化时需要备份。

4. 参数文件

数据库实例启动时需要读取参数文件,参数文件丢失可通过手工编辑一个,但修改过的参数将不能完全复原,建议参数修改后做备份。

5. 口令文件

口令文件记录的具有 SYSDBA 或 SYSOPER 权限用户的口令,如果丢失可重新创建。

6. 归档日志文件

归档文件是历史日志的内容复制品,在数据库恢复时需要使用,需要定期备份。

通过分析,上述 6 种文件中,需要定期备份的有数据文件、控制文件、参数文件和归档日志文件。

7.1.2 数据库备份

1. 非归档模式备份

非归档模式备份需要在关闭数据库的状态下进行文件备份,称为"冷备份"。首先关闭数据库,然后借助操作系统的复制命令进行文件复制,由于非归档模式下日志文件可能发生覆写,一旦发生覆写则旧备份将不能用于数据库完全恢复。

【例】 完成一次冷备份。

```
$ sqlplus / as sysdba
关闭数据库
SQL> shutdown immediate
Database closed.
Database dismounted.
ORACLE instance shut down.

物理备份
$ mkdir /u01/app/oradata/backup
备份数据文件
$ cp /u01/app/oradata/mydb/*.dbf   /u01/app/oradata/backup
备份控制文件
$ cp /u01/app/oradata/mydb/*.ctl   /u01/app/oradata/backup
备份日志文件
$ cp /u01/app/oradata/mydb/*.log   /u01/app/oradata/backup
备份参数文件
$ cp $ORACLE_HOME/dbs/spfile*.ora   /u01/app/oradata/backup
```

2. 归档模式备份

归档模式下可在数据库打开状态下进行文件备份,这种备份称为"热备份",当然也可执行冷备份。热备份过程中首先将数据库或表空间设置成备份状态,然后进行物理文件备份,备份完成后结束数据库或表空间的备份状态。

【例】 完成一次热备份。

```
$ sqlplus / as sysdba
设置归档模式
SQL> shutdown immediate
SQL> startup mount

SQL> alter database archivelog;
Database altered.

SQL> select log_mode from v$database;
```

```
LOG_MODE
--------------
ARCHIVELOG

SQL> alter database open;
```

完成一次热备份

备份数据文件
```
SQL> alter database begin backup;    -- 开始热备份
$ cp /u01/app/oradata/mydb/*.dbf   /u01/app/oradata/backup
SQL> alter database end backup;      -- 结束热备份
```

备份参数文件
```
$ cp $ORACLE_HOME/dbs/spfile*.ora  /u01/app/oradata/backup
```

备份控制文件
```
$ sqlplus / as sysdba
SQL> alter database backup controlfile to
    '/u01/app/oradata/backup/control.ctl'
```

备份归档日志
```
SQL> alter system archive log current;
SQL> quit
$ cp /u01/app/flash_recovery_area/…/*.ARC  /u01/app/oradata/backup
```

7.1.3 数据库恢复

恢复是指利用旧的数据库备份文件,在备份文件基础上执行 REDO 记录内容,将数据恢复成指定状态的动作。如果恢复到数据库故障点,称为完全恢复,数据没有任何损失,如果恢复到故障之前的某个点,称为不完全恢复,这时数据有一定损失。

1. 完全恢复

完全恢复是数据库恢复中最常用的,指利用备份文件和 REDO 文件将单个文件或单个表空间或整个数据库恢复到最新状态。执行数据库恢复前需要将备份文件以及归档日志还原到指定位置,然后执行 RECOVER 命令完成恢复。

恢复命令语法如下:

```
RECOVER
{ DATABASE
    | DATAFILE { filename |file_id
    | TABLESPACE tablespace }
```

属性说明:
1) DATABASE
指定恢复整个数据库,适用于有多个表空间或多个文件需要恢复的情况。

2) DATAFILE

指定要恢复的数据文件,可以是文件名或文件编号,如果有多个文件需要恢复用","隔开。

3) TABLESPACE

指定要恢复的表空间名。

【例】 完成数据文件恢复。

```
步骤一:关闭数据库备份数据库

步骤二:模拟数据变化
$ sqlplus / as sysdba
SQL > create table t01(id number) tablespace users;
SQL > insert into t01 values(100);
SQL > commit;

步骤三:模拟数据文件损坏
关闭数据库
删除表空间 users 对应的磁盘数据文件

步骤四:启动数据库
SQL > startup
ORACLE instance started.
Total System Global Area   313860096 bytes
Fixed Size                   1336232 bytes
Variable Size              201329752 bytes
Database Buffers           104857600 bytes
Redo Buffers                 6336512 bytes
Database mounted.
ORA - 01157: cannot identify/lock data file 4 - see DBWR trace file
ORA - 01110: data file 4: '/u01/app/oradata/mydb/users01.dbf'

步骤五:还原备份文件 USERS01.dbf
将备份的 USERS01.dbf 文件复制回原文件所在路径

步骤六:执行恢复
SQL > recover datafile 4;
Media recovery complete.

步骤七:打开数据库,验证表 t01 中数据
SQL > alter database open;
SQL > select * from t01;
ID
------
100
```

【例】 完成控制文件恢复。

```
步骤一：关闭数据库备份数据库

步骤二：模拟数据变化
$ sqlplus / as sysdba
SQL> create table t01(id number) tablespace users;
SQL> insert into t01 values(100);
SQL> commit;

步骤三：模拟控制文件损坏
关闭数据库
删除当前的控制文件并还原备份的控制文件

步骤四：启动数据库到 MOUNT 状态
SQL> startup mount
ORACLE instance started.
Total System Global Area   313860096 bytes
Fixed Size                   1336232 bytes
Variable Size              201329752 bytes
Database Buffers           104857600 bytes
Redo Buffers                 6336512 bytes
Database mounted.

步骤五：执行恢复
SQL> recover database using backup controlfile;

步骤六：打开数据库,验证表 t01 中数据
SQL> alter database open resetlogs
SQL> select * from t01;
ID
------
100
```

2. 不完全恢复

不完全恢复指利用备份文件和 REDO 文件将整个数据库恢复至过去的某个时点,数据有损失。

不完全恢复语法如下：

```
RECOVER DATABASE
{ UNTIL cancel
  | UNTIL TIME date
  | UNTIL CHANGE integer }
```

属性说明：
1) UNTIL cancel

指定基于用户取消类型的恢复,直到用户取消恢复操作。常用场景是恢复过程中归档文件部分丢失,这时只能恢复连续的归档文件,丢失后的归档将不再恢复。

2) UNTIL TIME date

指定基于时间的恢复,数据库只恢复到指定的时间点前。

3) UNTIL CHANGE integer

指定基于 SCN 号的恢复,数据库只恢复到指定的 SCN 号。

【例】 完成一次数据库不完全恢复。

```
步骤一：关闭数据库备份数据库

步骤二：模拟数据变化
$ sqlplus / as sysdba
SQL> create table t01(id number) tablespace users;
SQL> insert into t01 values(100);
SQL> commit;
SQL> select dbms_flashback.get_system_change_number scn from dual;
SCN
------
100023

SQL> insert into t01 values(200);
SQL> commit;
SQL> select dbms_flashback.get_system_change_number scn from dual;
SCN
------
100038

步骤三：还原所有数据文件
关闭数据库
将所有备份的数据文件复制回原来位置

步骤四：启动数据库到 MOUNT 状态
SQL> startup
ORACLE instance started.
Total System Global Area    313860096 bytes
Fixed Size                    1336232 bytes
Variable Size               201329752 bytes
Database Buffers            104857600 bytes
Redo Buffers                  6336512 bytes
Database mounted.

步骤五：执行不完全恢复
SQL> recover database until change 100023;

步骤六：打开数据库,验证表 t01 中数据
SQL> alter database open resetlogs;
SQL> select * from t01;
ID
------
100
```

数据库不完全恢复后必须指定 RESETLOGS 方式来打开数据库,重置当前日志序列号为 1,并归档以前尚未归档的日志。

7.2 恢复管理器

7.2.1 恢复管理器介绍

恢复管理器,就是熟称的 RMAN(Recovery Manager),实现一种服务器管理恢复(Server-Managed Recovery,SMR),SMR 是 Oracle 数据库通过执行数据库内置的备份恢复程序以确保成功备份恢复的能力。

RMAN 是 SMR 的具体实现,是 Oracle 数据库集成的备份和恢复工具,负责建立到 Oracle 数据库的连接,解析用户输入的备份恢复命令并调用数据库内部的备份和恢复包程序,RMAN 只负责命令解析,真正的代码执行由数据库上的进程来完成。RMAN 实用程序有两个组成部分:一是可执行命令,二是解析文件 recover.bsq,可执行命令通过读取解析文件获得在目标库上的程序调用,二者要求版本统一,否则不能正常工作。

借助操作系统命令是文件级备份,而使用 RMAN 进行备份是数据块级的备份,备份命令发出时,由备份通道进程将数据块从数据文件读入到输入缓冲区(通常位于 PGA 中),然后将多个文件输入缓冲区中需要备份的数据块复制至输出缓冲区,最后将输出缓冲区中的内容写出到磁盘备份文件,这个备份内容的集合称为备份集(Backup Set),对应的磁盘备份文件称为备份片(Backup Piece),备份过程如图 7-1 所示。

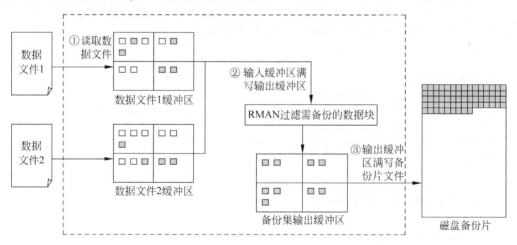

图 7-1 备份过程

输入缓冲区分配规则如下。

(1) 备份集文件数小于等于 4 个,每个文件分配 4 个 1MB 缓冲区,总和小于等于 16MB。

(2) 备份集文件数大于 4 个小于等于 8 个,每个文件分配 4 个 512KB 缓冲区,总和小于等于 16MB。

(3) 备份集文件数大于 8 个,每个文件分配 4 个 128KB 缓冲区。

输出缓冲区分配规则如下。
(1) 备份到磁盘,分配 4 个 1MB 缓冲区。
(2) 备份到磁带,分配 4 个 256KB 缓冲区。

利用 RMAN 进行数据还原时操作方式与备份类似,都是借助内存缓冲区,不同的是备份操作是读取数据文件并写入到备份文件,而还原操作是读取备份文件并写入数据文件。还原磁盘备份时,分配 4 个大小为 1MB 的输入缓冲区,还原磁带备份时,分配 4 个大小为 256KB 的输入缓冲区,用于还原的输出缓冲区为 4 个 128KB 大小的缓冲区。

使用 RMAN 进行备份,需要以 SYSDBA 身份连接数据库,RMAN 需要访问 SYS 用户的程序包,还要具有启动和关闭数据库的权限,但不像 SQL * Plus 工具那样要声明 SYSDBA 身份,基本语法如下:

```
RMAN TARGET target
```

其中,属性 TARGET 用来指定要备份的数据库,target 指定以 SYSDBA 身份连接要备份数据库的用户名/口令,可以使用操作系统验证"/",或者指定用户名/口令"sys/oracle",也可以通过网络服务名备份远程数据库"sys/oracle@remote"。

【例】 使用 RMAN 连接数据库。

```
在 Linux 系统本地连接
$ rman target /
Recovery Manager: Release 11.2.0.3.0 - Production on Wed Jul 17 20:53:47 2013
Copyright (c) 1982, 2011, Oracle and/or its affiliates.    All rights reserved.
connected to target database: MYDB (DBID=2726955368)
RMAN>

或远程客户端连接
C:\> rman target sys/oracle@remote
Recovery Manager: Release 11.2.0.1.0 - Production on Thu Jul 18 10:56:27 2013
Copyright (c) 1982, 2009, Oracle and/or its affiliates.    All rights reserved.
connected to target database: MYDB (DBID=2726955368)
RMAN>
```

【例】 使用 RMAN 备份数据库。

```
RMAN> backup database;
Starting backup at 18-JUL-13
using target database control file instead of recovery catalog

分配通道
allocated channel: ORA_DISK_1
channel ORA_DISK_1: SID=43 device type=DISK

指定备份文件,备份数据文件
```

```
channel ORA_DISK_1: starting full datafile backup set
channel ORA_DISK_1: specifying datafile(s) in backup set
input datafile file number = 00001 name = /u01/app/oradata/tsdb/system01.dbf
input datafile file number = 00002 name = /u01/app/oradata/tsdb/sysaux01.dbf
input datafile file number = 00003 name = /u01/app/oradata/tsdb/undotbs01.dbf
input datafile file number = 00005 name = /u01/app/oradata/tsdb/tbs.dbf
input datafile file number = 00004 name = /u01/app/oradata/tsdb/users01.dbf
channel ORA_DISK_1: starting piece 1 at 18 - JUL - 13
channel ORA_DISK_1: finished piece 1 at 18 - JUL - 13
piece handle = /u01/app/flash_recovery_area/TSDB/backupset/2013_07_18/o1_mf_nnndf_
TAG20130718T101936_8ygn5vsz_.bkp tag = TAG20130718T101936 comment = NONE
channel ORA_DISK_1: backup set complete, elapsed time: 00:00:35

指定备份文件,备份控制文件和参数文件
channel ORA_DISK_1: starting full datafile backup set
channel ORA_DISK_1: specifying datafile(s) in backup set
including current control file in backup set
including current SPFILE in backup set
channel ORA_DISK_1: starting piece 1 at 18 - JUL - 13
channel ORA_DISK_1: finished piece 1 at 18 - JUL - 13
piece handle = /u01/app/flash_recovery_area/TSDB/backupset/2013_07_18/o1_mf_ncsnf_
TAG20130718T101936_8ygn6ymp_.bkp tag = TAG20130718T101936 comment = NONE
channel ORA_DISK_1: backup set complete, elapsed time: 00:00:01
Finished backup at 18 - JUL - 13
```

7.2.2 RMAN 命令介绍

1. BACKUP

RMAN 使用 backup 命令执行备份,可备份数据库、表空间、数据文件、参数文件、控制文件以及归档日志,语法如下:

```
BACKUP
[ CHECK LOGICAL
  | CUMULATIVE
  | DURATION duration [ MINIMIZE TIME | MINIMIZE LOAD]
  | FILESPERSET integer
  | FORMAT format
  | TO DESTINATION destination
  | INCREMENTAL LEVEL integer
  | NOCHECKSUM
  | SECTION SIZE integer
  | VALIDATE ]
{ DATABASE
  | DATAFILE datafile
  | TABLESPACE tablespace
  | CURRENT CONTROLFILE
  | SPFILE
  | ARCHIVELOG ALL }
[ PLUS ARCHIVELOG ]
```

属性说明参见表 7-1。

表 7-1 backup 命令属性说明

参　数	描　述	范　例
CHECK LOGICAL	指定备份时执行逻辑检查	RMAN>backup check logical database
CUMULATIVE	指定累积备份，备份自上次增量 0 级备份后修改的数据块	RMAN>backup cumulative database
DURATION	指定备份最长可用时间，在规定时间内，可指定 MINIMIZE TIME 使备份尽可能快，或指定 MINIMIZE LOAD 使数据库负载尽可能小	RMAN>backup duration 00:10 minimize load
FILESPERSET	指定每个备份集可包含文件数	RMAN>backup filesperset 3 database
FORMAT	指定备份文件名格式。%d 指定数据库名，%T 指定日期，%U 指定系统产生唯一标识符	RMAN>backup format '%d_%T_%U.bkp' database
TO DESTINATION	指定备份文件存储路径	RMAN>backup to destination '/u01/bak' database;
INCREMENTAL LEVEL	指定增量备份，0 级代表全备份，1 级备份上次 0 级或 1 级备份后变化的数据	RMAN>backup incremental Level 0 database;
NOCHECKSUM	默认情况下，备份时会计算块的校验码存入块中，指定该属性将不计算	RMAN>backup nochecksum database;
SECTION SIZE	备份大文件时可分节进行备份	RMAN>backup section 300M datafile 5;
VALIDATE	只校验要备份的文件，不执行备份	RMAN>backup validate database;
DATABASE	指定全库备份	RMAN>backup database;
DATAFILE	指定要备份的文件	RMAN>backup datafile 4;
TABLESPACE	指定要备份的表空间	RMAN>backup tablespace users;
CURRENT CONTROLFILE	指定备份当前控制文件	RMAN>backup current controlfile;
SPFILE	指定备份参数文件	RMAN>backup spfile;
ARCHIVELOG ALL	指定备份归档	RMAN>backup archivelog all;
PLUS ARCHIVELOG	在备份数据库时可指定同时备份归档	RMAN>backup database plus archivelog;

【例】 备份表空间 users。

```
RMAN> backup tablespace users;

Starting backup at 18-JUL-13
using channel ORA_DISK_1
channel ORA_DISK_1: starting full datafile backup set
channel ORA_DISK_1: specifying datafile(s) in backup set
input datafile file number=00004 name=/u01/app/oradata/tsdb/users01.dbf
channel ORA_DISK_1: starting piece 1 at 18-JUL-13
channel ORA_DISK_1: finished piece 1 at 18-JUL-13
piece handle=/u01/app/flash_recovery_area/TSDB/backupset/2013_07_18/o1_mf_nnndf_
TAG20130718T141928_8yh27jo8_.bkp tag=TAG20130718T141928 comment=NONE
channel ORA_DISK_1: backup set complete, elapsed time: 00:00:01
Finished backup at 18-JUL-13
```

【例】 备份归档日志。

```
RMAN> backup archivelog all;

Starting backup at 18-JUL-13
current log archived
using channel ORA_DISK_1
channel ORA_DISK_1: starting archived log backup set
channel ORA_DISK_1: specifying archived log(s) in backup set
input archived log thread=1 sequence=16 RECID=1 STAMP=821110821
channel ORA_DISK_1: starting piece 1 at 18-JUL-13
channel ORA_DISK_1: finished piece 1 at 18-JUL-13
piece handle=/u01/app/flash_recovery_area/TSDB/backupset/2013_07_18/o1_mf_annnn_
TAG20130718T142022_8yh296vm_.bkp tag=TAG20130718T142022 comment=NONE
channel ORA_DISK_1: backup set complete, elapsed time: 00:00:03
Finished backup at 18-JUL-13
```

【例】 校验文件。

```
RMAN> backup validate database;

Starting backup at 18-JUL-13
using channel ORA_DISK_1
channel ORA_DISK_1: starting full datafile backup set
channel ORA_DISK_1: specifying datafile(s) in backup set
input datafile file number=00001 name=/u01/app/oradata/tsdb/system01.dbf
input datafile file number=00002 name=/u01/app/oradata/tsdb/sysaux01.dbf
input datafile file number=00003 name=/u01/app/oradata/tsdb/undotbs01.dbf
input datafile file number=00005 name=/u01/app/oradata/tsdb/tbs.dbf
input datafile file number=00004 name=/u01/app/oradata/tsdb/users01.dbf
channel ORA_DISK_1: backup set complete, elapsed time: 00:00:25
List of Datafiles
```

```
==================
File   Status   Marked Corrupt    Empty Blocks    Blocks Examined    High SCN
----   ------   --------------    ------------    ---------------    --------
1      OK       0                 16839           38400              335443
       File Name: /u01/app/oradata/tsdb/system01.dbf
       Block Type    Blocks Failing    Blocks Processed
       ----------    --------------    ----------------
       Data          0                 14524
       Index         0                 4591
       Other         0                 2446

File   Status   Marked Corrupt    Empty Blocks    Blocks Examined    High SCN
----   ------   --------------    ------------    ---------------    --------
2      OK       0                 31393           38400              335696
       File Name: /u01/app/oradata/tsdb/sysaux01.dbf
       Block Type    Blocks Failing    Blocks Processed
       ----------    --------------    ----------------
       Data 0        2083
       Index0        1686
       Other0        3238

File   Status   Marked Corrupt    Empty Blocks    Blocks Examined    High SCN
----   ------   --------------    ------------    ---------------    --------
3      OK       0                 129             23040              335696
       File Name: /u01/app/oradata/tsdb/undotbs01.dbf
       Block Type    Blocks Failing    Blocks Processed
       ----------    --------------    ----------------
       Data 0        0
       Index0        0
       Other0        22911

File   Status   Marked Corrupt    Empty Blocks    Blocks Examined    High SCN
----   ------   --------------    ------------    ---------------    --------
4      OK       0                 513             640                14181
       File Name: /u01/app/oradata/tsdb/users01.dbf
       Block Type    Blocks Failing    Blocks Processed
       ----------    --------------    ----------------
       Data          0                 0
       Index         0                 0
       Other         0                 127

File   Status   Marked Corrupt    Empty Blocks    Blocks Examined    High SCN
----   ------   --------------    ------------    ---------------    --------
5      OK       0                 1412            2560               279618
       File Name: /u01/app/oradata/tsdb/tbs.dbf
       Block Type    Blocks Failing    Blocks Processed
       ----------    --------------    ----------------
       Data          0                 86
       Index         0                 165
       Other         0                 897
```

```
channel ORA_DISK_1: starting full datafile backup set
channel ORA_DISK_1: specifying datafile(s) in backup set
including current control file in backup set
including current SPFILE in backup set
channel ORA_DISK_1: backup set complete, elapsed time: 00:00:01
List of Control File and SPFILE
===============================
File Type          Status   Blocks Failing   Blocks Examined
------------       ------   --------------   ---------------
SPFILE             OK       0                2
Control File       OK       0                594
Finished backup at 18-JUL-13
```

2. LIST

RMAN 使用 list 命令查看备份信息,语法如下:

```
LIST  [ EXPIRED ] BACKUP
[ OF {
        DATABASE
      | DATAFILE datafile
      | TABLESPACE tablespace
      | SPFILE
      | CONTROLFILE
      | ARCHIVELOG ALL } ]
[ BY FILE | SUMMARY ]
```

属性说明参见表 7-2。

表 7-2 list 命令属性说明

参 数	描 述	范 例
EXPIRED	显示失效的备份,当备份信息与物理备份文件不能匹配时即为失效,比如不小心把备份文件或归档日志删除	RMAN>list expired backup;
DATABASE	显示全库备份信息	RMAN>list backup of database;
DATAFILE	显示数据文件备份信息	RMAN>list backup of datafile 4;
TABLESPACE	显示表空间备份信息	RMAN > list backup of tablespace users;
SPFILE	显示参数文件备份信息	RMAN>list backup of spfile;
CONTROLFILE	显示控制文件备份信息	RMAN>list backup of controlfile;
ARCHIVELOG ALL	显示归档日志备份信息	RMAN>list backup of archivelog all;
BY FILE	根据文件分别显示备份信息	RMAN>list backup by file;
SUMMARY	显示汇总的备份信息	RMAN>list backup by summary;

【例】 显示汇总备份信息。

```
RMAN> list backup summary;

List of Backups
===============
Key     TY  LV  S  Device Type  Completion Time  #Pieces  #Copies  Compressed  Tag
-----   --  --  -  -----------  ---------------  -------  -------  ----------  ---
28      B   F   A  DISK         18-JUL-13        1        1        NO          TAG20130718T143626
29      B   F   A  DISK         18-JUL-13        1        1        NO          TAG20130718T143626
30      B   F   A  DISK         18-JUL-13        1        1        NO          TAG20130718T143951
```

【例】 显示数据文件备份信息。

```
RMAN> list backup of datafile 4;

List of Backup Sets
===================
BS Key   Type   LV   Size       Device Type   Elapsed Time   Completion Time
-------  ----   --   --------   -----------   ------------   ---------------
28       Full        253.16M    DISK          00:00:14       18-JUL-13

  List of Datafiles in backup set 28
  File  LV  Type  Ckp SCN    Ckp Time    Name
  ----  --  ----  --------   ---------   ----
  4         Full  336124     18-JUL-13   /u01/app/oradata/tsdb/users01.dbf

BS Key   Type   LV   Size       Device Type   Elapsed Time   Completion Time
-------  ----   --   --------   -----------   ------------   ---------------
30       Full        1.03M      DISK          00:00:00       18-JUL-13

  List of Datafiles in backup set 30
  File  LV  Type  Ckp SCN    Ckp Time    Name
  ----  --  ----  --------   ---------   ----
  4         Full  336124     18-JUL-13   /u01/app/oradata/tsdb/users01.dbf
```

【例】 显示文件分类备份信息。

```
RMAN> list backup by file;
List of Datafile Backups
========================
```

```
File  Key    TY  LV  S   Ckp SCN    Ckp Time     #Pieces  #Copies  Compressed  Tag
----  ----   --  --  --  --------   ---------    -------  -------  ----------  ---
1     28     B   F   A   336124     18-JUL-13    1        1        NO          TAG20130718T143626
2     28     B   F   A   336124     18-JUL-13    1        1        NO          TAG20130718T143626
3     28     B   F   A   336124     18-JUL-13    1        1        NO          TAG20130718T143626
4     30     B   F   A   336230     18-JUL-13    1        1        NO          TAG20130718T143951
      28     B   F   A   336124     18-JUL-13    1        1        NO          TAG20130718T143626
5     28     B   F   A   336124     18-JUL-13    1        1        NO          TAG20130718T143626

List of Control File Backups
============================

CF Ckp SCN     Ckp Time       BS Key   S   #Pieces  #Copies  Compressed  Tag
----------     --------       ------   --  -------  -------  ----------  ---
336130         18-JUL-13      29       A   1        1        NO          TAG20130718T143626

List of SPFILE Backups
======================

Modification Time       BS Key   S   #Pieces  #Copies  Compressed  Tag
-----------------       ------   --  -------  -------  ----------  ---
18-JUL-13               29       A   1        1        NO          TAG20130718T143626
```

3. REPORT

RMAN 使用 report 命令查看备份的分析信息,语法如下:

```
REPORT
{ OBSOLETE
  | NEED BACKUP DAYS integer }
```

属性说明参见表 7-3。

表 7-3 report 命令属性说明

参 数	描 述	范 例
OBSOLETE	报告废弃的备份	RMAN＞report obsolete;
NEED BACKUP DAYS	报告需要指定天数以上归档日志恢复的数据文件	RMAN＞report need backup days 3;

【例】 报告废弃备份信息。

```
RMAN> report obsolete;

RMAN retention policy will be applied to the command
RMAN retention policy is set to redundancy 1
Report of obsolete backups and copies
Type         Key     Completion Time   Filename/Handle
--------     ----    ---------------   ---------------
Archive Log  1       18-JUL-13         /u01/app/flash_recovery_area/
TSDB/archivelog/2013_07_18/o1_mf_1_16_8yh294on_.arc
```

【例】 报告需要备份文件。

```
RMAN> report need backup days 2;

Report of files whose recovery needs more than 2 days of archived logs
File Days Name
---- ---- -----------------------------------
1    83   /u01/app/oradata/tsdb/system01.dbf
2    83   /u01/app/oradata/tsdb/sysaux01.dbf
3    83   /u01/app/oradata/tsdb/undotbs01.dbf
4    83   /u01/app/oradata/tsdb/users01.dbf
5    83   /u01/app/oradata/tsdb/tbs.dbf
```

4. RESTORE

RMAN 使用 restore 命令还原备份文件,语法如下:

```
RESTORE
{ DATABASE
 | DATAFILE datafile
 | TABLESPACE tablespace
 | SPFILE
 | CONTROLFILE
 | ARCHIVELOG ALL }
[ FROM TAG tag
 | FROM AUTOBACKUP
 | FROM filename ]
[ PREVIEW ]
[ UNTIL SCN integer
 | UNTIL SEQUENCE integer
 | UNTIL TIME date ]
```

属性说明参见表 7-4。

表 7-4 restore 命令属性说明

参 数	描 述	范 例
DATABASE	还原数据库	RMAN>restore database;
DATAFILE	还原指定数据文件	RMAN>restore datafile 4;
TABLESPACE	还原指定表空间	RMAN>restore tablespace users;
SPFILE	还原参数文件	RMAN>restore spfile;
CONTROLFILE	还原控制文件	RMAN>restore controlfile;
ARCHIVELOG ALL	还原归档日志	RMAN>restore archivelog all;
FROM TAG	从指定标记的备份还原	RMAN>restore database from tag 'xxxx';
FROM AUTOBACKUP	从自动备份还原	RMAN > restore spfile from autobackup;
FROM	从指定文件还原	RMAN>restore spfile from '/u01/app/xxxxx.bkp'

续表

参　　数	描　　述	范　　例
PREVIEW	还原预览,不真正还原,主要检查备份是否能满足恢复	RMAN>restore database preview
UNTIL SCN	根据指定点还原备份	RMAN > restore database until scn 11000;
UNTIL SEQUENCE	根据指定日志序列号还原备份	RMAN > restore database until sequence 10;
UNTIL TIME	根据指定时间还原备份	RMAN>restore database until time TO_DATE('13-07-15','yy-mm-dd');

【例】 还原表空间 users。

```
RMAN > sql 'alter tablespace users offline';
sql statement: alter tablespace users offline

RMAN > restore tablespace users;
Starting restore at 18 – JUL – 13
using channel ORA_DISK_1

channel ORA_DISK_1: starting datafile backup set restore
channel ORA_DISK_1: specifying datafile(s) to restore from backup set
channel ORA_DISK_1: restoring datafile 00004 to /u01/app/oradata/tsdb/ users01.dbf
channel ORA_DISK_1: reading from backup piece /u01/app/ flash_recovery_area/ TSDB/backupset/
2013_07_18/o1_mf_nnndf_TAG20130718T152420_8yh6159b_.bkp
channel ORA_DISK_1: piece handle = /u01/app/flash_recovery_area/TSDB/
backupset/2013 _ 07 _ 18/o1 _ mf _ nnndf _ TAG20130718T152420 _ 8yh6159b _ . bkp tag
= TAG20130718T152420
channel ORA_DISK_1: restored backup piece 1
channel ORA_DISK_1: restore complete, elapsed time: 00:00:01
Finished restore at 18 – JUL – 13
```

5. RECOVER

RMAN 使用 recover 命令执行恢复,恢复过程中自动应用归档,语法如下:

```
RECOVER
{ DATABASE
 | DATAFILE datafile
 | TABLESPACE tablespace
 }
[ UNTIL SCN integer
 | UNTIL SEQUENCE integer
 | UNTIL TIME date ]
```

属性说明参见表 7-5。

表 7-5 recover 命令属性说明

参数	描述	范例
DATABASE	恢复数据库	RMAN＞recover database;
DATAFILE	恢复指定数据文件	RMAN＞recober datafile 4;
TABLESPACE	恢复指定表空间	RMAN＞recover tablespace users;
UNTIL SCN	根据指定点还原备份	RMAN＞recover database until scn 11000;
UNTIL SEQUENCE	根据指定日志序列号还原备份	RMAN＞recover database until sequence 10;
UNTIL TIME	根据指定时间还原备份	RMAN＞recover database until time TO_DATE('13-07-15', 'yy-mm-dd');

【例】 恢复表空间 users。

```
RMAN> sql 'alter tablespace users offline';
sql statement: alter tablespace users offline

RMAN> restore tablespace users;
Starting restore at 18－JUL－13
using channel ORA_DISK_1

channel ORA_DISK_1: starting datafile backup set restore
channel ORA_DISK_1: specifying datafile(s) to restore from backup set
channel ORA_DISK_1: restoring datafile 00004 to /u01/app/oradata/tsdb/users01.dbf
channel ORA_DISK_1: reading from backup piece /u01/app/flash_recovery_area/TSDB/backupset/2013_07_18/o1_mf_nnndf_TAG20130718T152420_8yh6159b_.bkp
channel ORA_DISK_1: piece handle = /u01/app/flash_recovery_area/TSDB/backupset/2013_07_18/o1_mf_nnndf_TAG20130718T152420_8yh6159b_.bkp tag = TAG20130718T152420
channel ORA_DISK_1: restored backup piece 1
channel ORA_DISK_1: restore complete, elapsed time: 00:00:01
Finished restore at 18－JUL－13

RMAN> recover tablespace users;
Starting recover at 18－JUL－13
using channel ORA_DISK_1
starting media recovery
media recovery complete, elapsed time: 00:00:01
Finished recover at 18－JUL－13

RMAN> sql 'alter tablespace users online';
sql statement: alter tablespace users online
```

6. DELETE

RMAN 使用 delete 命令删除备份，语法如下：

```
DELETE
[ NOPROMPT ]
[ OBSOLETE ]
[ EXPIRED ]
BACKUP
```

属性说明参见表 7-6。

表 7-6 delete 命令属性说明

参　数	描　述	范　例
NOPROMPT	指定删除时无提示	RMAN>delete noprompt backup;
OBSOLETE	指定删除废弃的备份	RMAN>delete obsolete;
EXPIRED	指定删除失效的备份	RMAN>delete expired backup;
BACKUP	指定删除备份	RMAN>delete backup;

【例】 删除废弃备份。

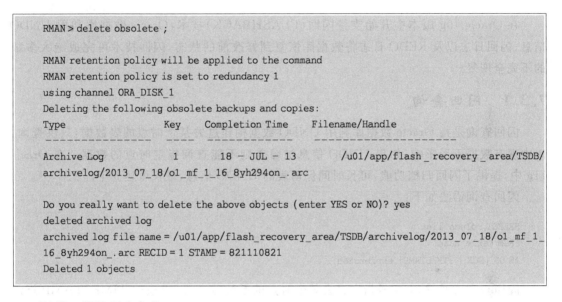

【例】 删除所有备份。

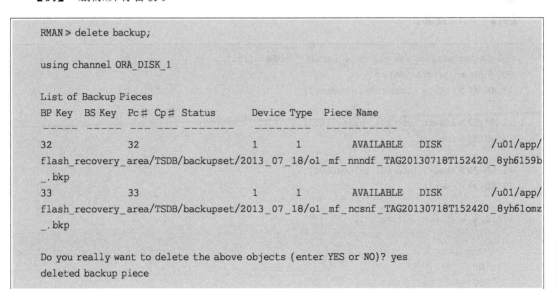

```
backup piece handle = /u01/app/flash_recovery_area/TSDB/backupset/2013_07_18/o1_mf_nnndf_
TAG20130718T152420_8yh6159b_.bkp RECID = 32 STAMP = 821114661
deleted backup piece
backup piece handle = /u01/app/flash_recovery_area/TSDB/backupset/2013_07_18/o1_mf_ncsnf_
TAG20130718T152420_8yh61omz_.bkp RECID = 33 STAMP = 821114677
Deleted 2 objects
```

7.3 闪回技术

在 Oracle 10g 版本中开始支持闪回(FLASHBACK)技术,Oracle 数据库利用 UNDO 信息、闪回日志以及 REDO 日志将数据库恢复到修改前的状态,闪回技术可完成绝大多数的不完全恢复。

7.3.1 闪回查询

闪回查询是指 Oracle 数据库利用 UNDO 数据构建过去某个时点的表数据,只供查询,表中现有数据不受影响,如果 UNDO 信息被覆盖则不能查询指定时点的数据。在 Oracle 11g 中,提供了闪回归档功能,可长时间保留表的 UNDO 信息,最长可达 5 年。

闪回查询语法如下:

```
SELECT column_list
FROM table_name
AS OF {SCN | TIMESTAMP} expression;
```

闪回查询可通过 SCN 号指定过去某个时点的系统改变号,或通过 TIMESTAMP 指定过去某个时间点。

【例】 闪回查询。

```
SQL> alter session set nls_date_format = 'hh24:mi:ss';
SQL> set sqlprompt _date>
13:48:41> create table t01(id number);

13:48:52> insert into t01 values(100);
13:48:55> commit;

13:50:35> insert into t01 values(200);
13:50:39> commit;

13:50:41> select * from t01;
    ID
--------
   100
   200
13:51:37> select * from t01
```

```
            as of timestamp
            to_timestamp('2013/09/20 13:49:30','yyyy/mm/dd hh24:mi:ss');
    ID
----------
   100
```

7.3.2 闪回数据

闪回数据是指 Oracle 数据库利用 UNDO 数据构建过去某个时点的表数据，如果 UNDO 信息被覆盖则不能构建指定时点的数据。闪回表数据前需要修改表以支持行数据移动。

闪回数据语法如下：

```
FLASHBACK TABLE table_name
TO {SCN | TIMESTAMP} expression;
```

【例】 闪回数据。

```
13:48:52> insert into t01 values(100);
13:48:55> commit;

13:50:35> insert into t01 values(200);
13:50:39> commit;

13:52:51> delete from t01;
13:52:58> commit;

13:53:42> select * from t01;
未选定行

13:55:45> alter table t01 enable row movement;
13:56:15> flashback table t01
         to timestamp
         to_timestamp('2013/09/20 13:50:00','yyyy/mm/dd hh24:mi:ss');
13:56:37> select * from t01;
    ID
----------
   100

13:56:15> flashback table t01
         to timestamp
         to_timestamp('2013/09/20 13:52:00','yyyy/mm/dd hh24:mi:ss');
13:56:37> select * from t01;
    ID
----------
   100
   200
```

7.3.3 闪回删除

当通过 DROP 语句删除表时,如未指定 PURGE 语句,表被改名放入回收站中,表占用的空间未释放,如果使用了 PURGE 语句,则表彻底删除。当表被改名放入回收站后,可通过查看回收站视图 USER_RECYCLEBIN 获得删除对象信息,并可通过闪回语句将删除的表恢复成删除前状态。

闪回删除语法如下:

```
FLASHBACK TABLE table_name
TO BEFORE DROP
[ RENAME TO new_table_name ];
```

闪回删除表时,如果表名已经被占用,可通过 RENAME TO 将闪回的表修改成新表名。

【例】 闪回删除。

```
14:12:40> select * from t01;
    ID
----------
   100
   200

14:12:42> drop table t01;
14:12:44> select * from t01;
select * from t01
              *
ERROR at line 1:
ORA-00942: table or view does not exist

14:12:46> select object_name,original_name from user_recyclebin;
OBJECT_NAME                          ORIGINAL_NAME
-----------------------------        -------------
BIN$g/dP+DBGRt2xPN5KjtTyNA==$0       T01

14:13:27> select * from "BIN$g/dP+DBGRt2xPN5KjtTyNA==$0";
    ID
----------
   100
   200

14:13:40> flashback table t01 to before drop;

14:13:52> select * from t01;
    ID
----------
   100
   200
```

7.3.4 闪回版本查询

闪回版本查询是指查询某段时间内表上数据的变化。闪回版本查询语法如下：

```
SELECT column_list,versions_column_list
FROM table_name
VERSIONS BETWEEN {SCN| TIMESTAMP} {MINVALUE | expression}
AND {MAXVALUE |expression}
```

versions_column_list 指定闪回版本查询时查询的伪列，Oracle 提供了 6 个伪列，见表 7-7。通过 VERSIONS 语句指定版本查询范围，可指定 SCN 范围，也可以指定 TIMESTAMP 返回。

表 7-7　闪回版本查询伪列

伪　　列	描　　述
VERSIONS_STARTSCN	行数据第一次使用时的 SCN 号
VERSIONS_ENDSCN	行数据版本过期的 SCN 号
VERSIONS_STARTTIME	行数据第一次使用时的时间戳
VERSIONS_ENDTIME	行数据版本过期的时间戳
VERSIONS_XID	创建执行版本的事务号
VERSIONS_OPERATION	执行事务的操作（I 表示插入，U 表示更新，D 表示删除）

【例】　闪回版本查询。

```
SQL> create table t02(id number,name varchar2(20));
SQL> insert into t02 values(1,'a');
SQL> commit;

SQL> insert into t02 values(2,'b');
SQL> update t02 set name = 'xx' where id = 1;
SQL> commit;

SQL> insert into t02 values(3,'c');
SQL> commit;

SQL> select versions_startscn,versions_endscn,versions_xid,
       versions_operation,id,name
       from t02
       versions between scn minvalue and maxvalue;
VERSIONS_STARTSCN  VERSIONS_ENDSCN  VERSIONS_XID      V   ID   NAME
-----------------  ---------------  ----------------  --  ---  ----
           320114                   02001600BD000000  I    3   c
           320109                   09000600CC000000  U    1   xx
           320109                   09000600CC000000  I    2   b
           320098           320109  04000400C4000000  I    1   a
```

7.3.5 闪回事务

闪回事务是指根据表上的事务变化（通过闪回版本查询获得），根据事务号查询闪回事务表 FLASHBACK_TRANSACTION_QUERY 获得事务对应的 UNDO_SQL，然后手动执行这些 SQL 来恢复指定事务。

Oracle 11g 包 DBMS_FLASHBACK 中新增方法 TRANSACTION_BACKOUT，该方法通过 LOGMNR 挖掘日志进行事务恢复。

【例】 使用 DBMS_FLASHBACK 包闪回事务。

```
SQL> select versions_startscn,versions_endscn,versions_xid,
    versions_operation,id,name
    from t02
    versions between scn minvalue and maxvalue;
VERSIONS_STARTSCN  VERSIONS_ENDSCN  VERSIONS_XID      V  ID  NAME
-----------------  ---------------  ----------------  -- --- ------
       320114                       02001600BD000000  I   3   c
       320109                       09000600CC000000  U   1   xx
       320109                       09000600CC000000  I   2   b
       320098           320109      04000400C4000000  I   1   a

以 SYS 用户登录进行事务恢复
declare
   xids sys.xid_array;
begin
   xids:= sys.xid_array('09000600CC000000');
   dbms_flashback.transaction_backout(
        numtxns =>1,
        xids    =>xids,
        options =>dbms_flashback.cascade);
end;

恢复后数据,第二个事务操作全部恢复
SQL> select * from u1.t02;
    ID NAME
------ --------------------
     1 a
     3 c
```

7.3.6 闪回数据库

闪回数据库可以说是另一种更高效的数据库不完全恢复方法，Oracle 数据库利用闪回日志进行数据库恢复。使用闪回数据库特性前需开启数据库闪回日志：alter database flashback on。同时通过系统参数 DB_FLASHBACK_RETENTION_TARGET 设置闪回日志的保留时间。

闪回数据库语法如下：

```
FLASHBACK DATABASE
TO {SCN| TIMESTAMP} expression;
```

【例】 闪回数据库。

```
SQL> shutdonw immediate
SQL> startup mount
SQL> select flashback_on from v$database;
FLASHBACK_ON
------------------
NO
SQL> alter database flashback on;
SQL> alter database open;

15:01:01> create table t03(id number);
15:01:05> insert into t03 values(100);
15:01:08> commit;

15:03:05> insert into t03 values(200);
15:03:08> commit;

闪回数据库
SQL> shutdown immediate
SQL> startup mount
SQL> flashback database to timestamp
     to_timestamp('2013/09/20 15:02:00', 'yyyy/mm/dd hh24:mi:ss');
SQL> alter database open resetlogs;
SQL> select * from t03;
    ID
---------
   100
```

7.3.7 闪回归档

闪回归档的本质是利用表空间存储表的数据变化,使变化的数据可以长久保留,不再受 UNDO 表空间的限制。使用闪回归档,需先创建存储变化数据的表空间,然后创建闪回归档对象,最后修改表使用闪回归档记录数据变化。

创建闪回归档语法如下:

```
CREATE FLASHBACK ARCHIVE flashback_archive_name
TABLESPACE tablespace_name
RETENTION n{day | month | year };
```

其中,TABLESPACE 语句指定闪回归档数据存储空间,RETENTION 指定保留时间,最长 5 年。

```
SQL> create tablespace fa datafile 'fs.dbf' size 20M autoextend on next 5M;
SQL> create flashback archive fa1 tablespace fa retention 5 month;
SQL> alter table t01 flashback archive fa1;
```

小　　结

本章主要讲解了数据库备份恢复和闪回技术。数据库备份恢复可以手动完成,也可借助工具完成,当前使用最多的是使用 RMAN 进行备份恢复。备份恢复部分重点掌握工具 RMAN 的使用,包括单个文件、单个表空间以及整个数据库的备份恢复。闪回技术是 Oracle 10g 开始提供的特性,主要包括闪回查询、闪回数据、闪回删除、闪回事务以及闪回数据库,Oracle 11g 新增了闪回归档可增大 UNDO 数据保留时长。

思　考　题

1. 数据库有哪些文件？备份数据库时哪些文件需要备份？
2. 假设数据文件 5 损坏,现有之前的完整备份和归档日志,写出恢复该文件的步骤。
3. 简述冷备份和热备份的概念。
4. 简述备份集和备份片概念及关联关系。
5. 假设数据库有完整备份,数据库异常关闭后参数文件、控制文件和数据文件全部损坏,在线日志文件、归档日志文件完好,写出完全恢复数据库步骤。
6. 闪回技术中,哪些需要借助 UNDO 数据完成？
7. 创建一个闪回归档 fa1,保留时间 3 年,并修改表 t01 使用该闪回归档,写出操作语句。

第三部分

Oracle数据库开发

第 8 章　Schema 对象

Schema 是数据库用户拥有的数据库对象的集合,包括表、索引、视图、序列等。本章将逐个介绍每个 Schema 对象的特性及应用场景。

8.1　表

8.1.1　数据类型

数据库中存储数据的类型主要包括字符型、数字型以及日期型,下面将介绍每种类型中常用的数据类型。

1. 字符数据类型

1) char(n[byte|char])

char 类型最长可存储 2000 个字节数据,默认存储单位是字节(byte),使用数据库字符集。如实际字符数据长度小于声明的最大长度,存储时不足的字节或字符以空格填充。

```
create table test (name char(2));          //可插入两个英文字母或一个汉字
create table test (name char(2char));      //可插入两个英文字母或两个汉字
```

如果定义为 name char(2000 char),实际上不能存储 2000 汉字,只能存储 1000 汉字,因为该类型最大存储 2000 字节,但可存储 2000 英文字符。

2) varchar2(n[byte|char])

varchar2 类型最长可存储 4000 个字节数据,默认存储单位是字节(byte),使用数据库字符集。如实际字符数据长度小于声明的最大长度,存储时分配空间以实际数据长度为准,该类型称作变长的数据类型。

3) nchar(n)

nchar 类型最长可存储 2000 个字节数据,使用国家字符集,这里的 n 可设多大要看数据库的国家字符集,国家字符集可以是 AL16UTF16 或 UTF8,假如国家字符集是 AL16UTF16,则无论什么字符都以两个字节存放,这时 n 最大为 1000。

4) nvarchar2(n)

nvarchar2 类型最长可存储 4000 个字节数据,使用国家字符集,这里的 n 可设多大要看数据库的国家字符集,国家字符集可以是 AL16UTF16 或 UTF8,假如国家字符集是 AL16UTF16,则无论什么字符都以两个字节存放,这时 n 最大为 2000。

2. 数字数据类型

1) number[(p[,s])]

p 指定精度,取值范围 1~38,指小数点左右数字位数,s 指定小数位数,取值范围 -84~127,如数据小数位数多于定义位数则进行四舍五入,number 类型数据占用空间最大 22 个字节,计算方法为 length=floor((p+1)/2)+1。

```
create table test1(nnumber(5,2));
//插入 123 可以,1234 不可以,因为 1234 是整数 4 位再加上两位小数共 6 位,超出精度 5
create table test2(n number(5,-2));
//插入 123456 可以,小数位数是负数,向小数点左侧进行四舍五入
insert into test2 values(123456);
select * from test2;
         N
----------
    123500
```

2) binary_float

二进制浮点数,用 32 位表达小数,存储占 4 个字节。单精度浮点数存储包含 8 位指数位、24 位尾数,另外还有一位隐藏符号位,精度为 $24 \times \log 102$,约等于 7。

```
create table test( n binary_float );
insert into test values(1.234512345);
insert into test values(1234512345);
select * from test;
         N
-----------------------------------------------------
         1.23451233
       1234512380.00000000
```

3) binary_double

二进制浮点数,用 64 位表达小数,存储占 8 个字节。双精度浮点数存储包含 11 位指数位、53 位尾数,另外还有一位隐藏符号位,精度为 $53 \times \log 102$,约等于 15。

```
create table test1( n binary_double );
insert into test1 values(1.234512345123458888);
insert into test1 values(12345123451234512345);
select * from test1;
         N
-----------------------------------------------------
               1.234512345123459
     12345123451234513000.000000000000000
```

3. 日期数据类型

1) date

用于存储日期和时间,包括年月日时分秒。

2) timestamp[(n)]

date 类型扩展,可存储秒的小数位,取值范围 1~9,默认为 6。

3) timestamp[(n)]with time zone

除存储基本年月日时分秒以及小数秒外,还存储时区信息。

4) timestamp[(n)]with local time zone

与 timestamp[(n)]with time zone 的不同之处为,存储日期时直接转换为数据库时区日期,而读取时将数据库时区的日期转换为用户会话时区日期。

```
create table date_test(
d1 timestamp with time zone,
d2 timestamp with local time zone);
insert into date_test values(systimestamp,systimestamp);
select * from date_test;
D1                                              D2
---------------------------------------------   ---------------------------
08-7月 -13 08.33.35.832723 上午 +08:00          08-7月 -13 08.33.35.83
```

5) interval year[(years_precision)] to month

时间间隔,单位为年到月,years_precision 指定年的精度,默认为 2。

```
create table dura(dura interval year(4) to month);
insert into dura values(interval '11' month);
insert into dura values(interval '1-3' year to month);
select * from dura;
DURA
---------------------------
+00-11
+01-03
```

6) interval day[(days_precision)] to second[(seconds_precision)]

时间间隔,单位为天到秒,days_precision 指定天的精度,默认为 2,seconds_precision 指定秒的精度,默认为 6。

```
create table dura(dura interval day(4) to second(4));
insert into dura values(interval '3' day);
insert into dura values(interval '3 2' day to hour);
insert into dura values(interval '2:30' hour to minute);
select * from dura;
DURA
---------------------------
+0003 00:00:00.0000
+0003 02:00:00.0000
+0000 02:30:00.0000
```

4. 大对象类型

1) clob

clob 用于存储大型的字符数据,使用数据库字符集,最大可达 4GB×数据库块大小,如果块为 8KB,最大可达 32TB,数据变长。

2) nclob

nclob 用于存储大型的字符数据,使用国家字符集,最大可达 4GB×数据库块大小,如果块为 8KB,最大可达 32TB,数据变长。

3) blob

blob 用于存储大型的二进制数据,最大可达 4GB×数据库块大小,如果块为 8KB,最大可达 32TB,数据变长。可存储图片、影音等数据。

4) bfile

存储文件指针,文件存储在数据库外。

8.1.2 堆表

堆表(heap table)就是人们常用的数据库表,表中存储的数据是无序的。

1. 创建表

创建表语法如下:

```
CREATE TABLE table_name(
column_name datatype
[, column_name datatype ]
)
[ SEGMENT CREATION { IMMEDIATE | DEFERRED } ]
[ TABLESPACE tablespace_name ]
[ LOGGING | NOLOGGING ]
[ PCTUSED integer ]
[ PCTFREE integer ]
STORAGE(
    INITIAL size
    MAXSIZE size
    FREELISTS integer
    BUFFER_POOL { default | keep | recycle }
);
```

属性说明:

1) table_name

指定要创建的表名称。

2) column_name datatype

指定表中列的名称及数据类型,如果存在多个列,则用","隔开。

3) SEGMENT CREATION

指定创建表时是否立即或延迟创建段。Oracle 11g 新增了一个参数用于控制段是否立即创建,参数 DEFERRED_SEGMENT_CREATEION 默认为 true,意味着建表时不创建段,当第一次插入数据时再分配段。如果建表时未指定段创建的语句,则默认采用系统设置。IMMEDIATE 指定立即创建段,DEFERRED 指定延迟创建段。

4) TABLESPACE

指定表存储于哪个表空间。如果不指定该语句,则表默认存储于用户的默认表空间中。

5) LOGGING | NOLOGGING

指定表的日志记录方式,LOGGING 指定记录日志,NOLOGGING 指定不记录日志。

6) PUTUSED

适用于手动段空间管理方式下,通过该参数设置数据块满后重新回到自由链表的控制线。integer 指定一个数字,比如 40 意味着数据块满以后块数据下降到块空间的 40% 以下方可再插入新数据。

7) PCTFREE

指定数据块的空间空间,默认为 10,也就是数据块在插入数据时要保留 10% 空间,一旦插入数据后空闲空间接近 10%,则该块不再允许插入新数据。

8) STORAGE

指定表数据存储的选项设置。

9) INITIAL

指定创建表时分配空间的大小。如果区大小是系统自动分配,则根据设置大小 Oracle 自动创建区,区大小可能是 64KB、1MB、8MB 和 64MB,如果该参数设置为 16MB,则 Oracle 会创建两个 8MB 区,如果该参数设置为 10MB,则 Oracle 会创建一个 8MB 区和两个 1MB 区。如果区大小是统一大小,比如 2MB,如果该参数设置为 4MB,则 Oracle 会创建两个区,如果该参数设置为 8MB,则 Oracle 会创建 4 个区。

10) MAXSIZE

指定数据表占用空间的最大值。

11) FREELIST

适用于手动段空间管理方式下,指定自由链表数量。

12) BUFFER_POOL

指定表数据加载进 SGA 时,存储的缓冲区,可以是默认的缓冲区 DEFAULT、保留池 KEEP 或回收池 RECYCLE。

【例】 创建学生表,包含信息:学号、姓名、年龄、生日。

```
create table student(
    sno number(8),
    sname varchar2(20),
    sage number(3),
    sbirth date
);
```

【例】 创建教师表,包含信息:教师编号、姓名、年龄、生日、个人简历,存储表空间 tbs1 中。

```
create table teacher(
    tno number(8),
    tname varchar2(20),
    tage number(3),
    tbirth date,
    tresume clob
)
tablespace tbs1;
```

【例】 创建院系表,包含信息:院系编号、名称,初始分配空间 4MB。

```
create table department(
   dno number(8),
   dname varchar2(20)
)
storage(initial 4M);
```

2. 修改表

修改表语法如下:

```
ALTER TABLE table_name
[ LOGGING | NOLOGGING ]
[ RENAME TO new_table_name ]
[ SHRINK SPACE ]
[ READ ONLY | READ WRITE ]
[ ADD (column_name datatype [, …] ) ]
[ MODIFY ( column_name datatype [, …] ) ]
[ DROP  {COLUMN column_name | ( column_name, [, …] ) ]
[ RENAME COLUMN old_name TO new_name ];
```

属性说明:

1) table_name

指定要修改的表名称。

2) LOGGING | NOLOGGING

指定表的日志记录方式,logging 指定记录日志,nologging 指定不记录日志。

3) RENAME TO

对表名进行修改。

4) SHRINK SPACE

收缩表,当表数据经过大量删除后,在不同块都会有大量空闲空间,该语句可将数据进行整理,然后将表占用空间进行收缩。

5) READ ONLY | READ WRITE

将表修改为只读或可读写。

6) ADD

对表进行列增加。

7) MODIFY

修改列的数据类型。

8) DROP

删除单个列或多个列。

9) RENAME COLUMN

修改列的名字。

【例】 创建表,对表进行修改。

```
SQL> create table t01(id number);
SQL> alter table t01 add (name varchar2(20));
SQL> alter table t01 modify name varchar2(30);
SQL> alter table t01 rename column name to username;
SQL> alter table t01 drop column username;
SQL> alter table t01 rename to emp;

SQL> alter table emp read only;
SQL> insert into emp values(100);
ERROR at line 1:
ORA-12081: update operation not allowed on table "U1"."EMP"

SQL> alter table emp read write;
SQL> insert into emp values(100);
1 row created.
```

3. 删除表

删除表语法如下：

```
DROP TABLE table_name
[ CASCADE CONSTRAINTS ]
[ PURGE ]
```

属性说明：

1) table_name

指定要删除的表名称。

2) CASCADE CONSTRAINTS

指定删除表时，删除所有引用该表的引用约束，如果不使用该语句，则当表被其他表的外键引用时不能删除表。

3) PURGE

指定删除表时同时释放空间。

8.1.3 临时表

临时表(global temporary table)中的数据都是临时存在，不能持久存储，多用于在数据统计时中间结果的存储。相同用户的不同会话间彼此不能共享数据。临时表的创建语法如下，表修改和删除与堆表类似。

创建临时表语法如下：

```
CREATE GLOBAL TEMPORARY TABLE table_name(
column_name datatype
[, column_name datatype ]
)
[ ON COMMIT { DELETE | PRESERVE } ROWS ];
```

属性说明：

1）table_name

指定要创建的表名称。

2）column_name datatype

指定表中列的名称及数据类型，如果存在多个列，则用","隔开。

3）ON COMMIT

指定临时表中数据在事务结束时数据的处理方式，ON COMMIT DELETE ROWS 是默认设置，该设置指定在事务结束时将临时表中数据删除，ON COMMIT PRESERVE ROWS 指定当事务结束时临时表数据保留，一直保留到会话结束。

【例】 创建临时表，默认提交删除。

```
SQL> create global temporary table temp1(id number);
SQL> insert into temp1 values(100);

SQL> select * from temp1;
ID
----------
100

SQL> commit;
SQL> select * from temp1;
no rows selected
```

【例】 创建临时表，设置提交保留数据。

```
SQL> create global temporary table temp1(id number) on commit preserve rows;
SQL> insert into temp1 values(100);
SQL> select * from temp1;
ID
----------
100

SQL> commit;
SQL> select * from temp1;
ID
----------
100
```

8.1.4　索引组织表

索引组织表（index-organized table）指以索引数据的存储方式来存储表数据，可大幅度提升表数据查询效率，但是如果表数据更改频繁，则需重新对数据排序导致额外开销，适用于数据查询频繁且数据很少变更的表。索引组织表必须有主键列，表数据按主键列排序。索引组织表的创建语法如下，表修改和删除与堆表类似。

创建索引组织表语法如下：

```
CREATE TABLE table_name(
column_name datatype
[, column_name datatype ]
)
ORGANIZATION INDEX;
```

属性说明：

1) table_name

指定要创建的表名称。

2) column_name datatype

指定表中列的名称及数据类型，如果存在多个列，则用","隔开。

3) ORGANIZATION INDEX

指定表数据组织方式为索引方式。

【例】 创建索引组织表。

```
create table city(
    id number primary key,
    name varchar2(20)
)
organization index;
```

8.1.5 分区表

分区表适用于表数据量很大的情况，根据分区列将数据分布到多个分区表中，可减缓单个表的访问及维护压力。表数据达到几百万行、几千万行甚至更高时可考虑使用分区表。Oracle 11g 对表分区技术做了大量改进，除原有的范围分区、列表分区、哈希分区外，新增了间隔分区、引用分区、系统分区，分区组合方式更加灵活。接下来主要介绍最常用的分区方式以及组合分区方式。

1. 创建表分区

1) 范围分区

语法：

```
CREATE TABLE table_name(
column_name datatype
[, column_name datatype ]
)
PARTITION BY RANGE ( column_name )
(
    PARTITION partition_name VALUES LESS THAN { literal | MAXVALUE }
    [, …]
)
```

属性说明：

(1) table_name

指定要创建的表名称。

(2) column_name datatype

指定表中列的名称及数据类型,如果存在多个列,则用","隔开。

(3) PARTITION BY RANGE

通过关键字 PARTITION BY RANGE 声明是范围分区,括号中 column_name 指定范围分区列。

(4) PARTITION

partition_name 指定分区名,literal 指定一个具体值用来限定范围。该语句的作用是定义值小于 literal 的在一个分区中,定义最后一个分区时指定值小于 MAXVALUE。

【例】 创建员工表,按员工编号分区,第一个分区员工号≤1000号,第二个分区为 1001~2000 号,其余放第三个分区。

```
create table emp(
    eid number primary key,
    ename varchar2(20),
    egender varchar2(10),
    eage number,
    ebirth date
)
prtition by range ( eid )
(
    partition p1 values less than(1001),
    partition p2 values less than(2001),
    partition p3 values less than(maxvalue)
);
```

【例】 创建员工表,按员工生日分区,第一个分区存生日在 1990 年 1 月 1 日之前的员工,第二个分区存生日在 1990 年 1 月 1 日~1999 年 12 月 31 日之间的员工,其余放第三个分区。

```
create table emp(
    eid number primary key,
    ename varchar2(20),
    egender varchar2(10),
    eage number,
    ebirth date
)
prtition by range ( ebirth )
(
    partition p1 values less than(to_date('1990-01-01','yyyy-mm-dd')),
    partition p2 values less than(to_date('2000-01-01','yyyy-mm-dd')),
    partition p3 values less than(maxvalue)
);
```

2) 列表分区

语法：

```
CREATE TABLE table_name(
column_name datatype
[, column_name datatype ]
)
PARTITION BY LIST ( column_name )
(
    PARTITION partition_name VALUES { literal | DEFAULT}
    [, … ]
)
```

属性说明：

(1) table_name

指定要创建的表名称。

(2) column_name datatype

指定表中列的名称及数据类型，如果存在多个列，则用","隔开。

(3) PARTITION BY LIST

通过关键字 PARTITION BY LIST 声明是列表分区，括号中 column_name 指定列表分区列。

(4) PARTITION

partition_name 指定分区名，literal 指定一个具体值用来限定范围。该语句的作用是定义值等于 literal 的在一个分区中，定义最后一个分区时指定值为 DEFAULT。

【例】 创建员工表，按员工性别分区，第一个分区存男性员工，第二个分区存女性员工，其余放第三个分区。

```
create table emp(
    eid number primary key,
    ename varchar2(20),
    egender varchar2(10),
    eage number,
    ebirth date
)
prtition by list ( egender )
(
    partition p1 values ('男'),
    partition p2 values ('女'),
    partition p3 values (DEFAULT)
);
```

3) 哈希分区

语法：

```
CREATE TABLE table_name(
column_name datatype
```

```
[, column_name datatype ]
)
PARTITION BY HASH ( column_name )
(
    PARTITION partition_name
    [, … ]
)
```

属性说明：

（1）table_name

指定要创建的表名称。

（2）column_name datatype

指定表中列的名称及数据类型，如果存在多个列，则用","隔开。

（3）PARTITION BY HASH

通过关键字 PARTITION BY HASH 声明是哈希分区，括号中 column_name 指定哈希分区列。

（4）PARTITION

partition_name 指定分区名。

【例】 创建员工表，按员工年龄分区，自动存入两个分区中。

```
create table emp(
    eid number primary key,
    ename varchar2(20),
    egender varchar2(10),
    eage number,
    ebirth date
)
prtition by hash( age )
(
    partition p1,
    partition p2
);
```

4）间隔分区

间隔分区适用于范围分区，只支持数值型和日期型。分区时只需指定第一个分区，其余分区由 Oracle 数据库根据插入值范围自动创建相应分区。

语法：

```
CREATE TABLE table_name(
column_name datatype
[, column_name datatype ]
)
PARTITION BY RANGE ( column_name ) INTERVAL ( literal )
(
    PARTITION partition_name VALUES LESS THAN (literal )
)
```

属性说明：

(1) table_name

指定要创建的表名称。

(2) column_name datatype

指定表中列的名称及数据类型，如果存在多个列，则用","隔开。

(3) PARTITION BY RANGE (column_name) INTERVAL(literal)

通过关键字 PARTITION BY RANGE 声明是范围分区，括号中 column_name 指定范围分区列。literal 指定间隔值，Oracle 数据库根据插入值自动创建相应分区。

(4) PARTITION

partition_name 指定分区名，literal 指定一个具体值用来限定范围。该语句定义第一个分区。

【例】 创建员工表，按员工编号分区，第一个分区员工号≤1000号，之后每隔1000号建一个分区。

```
create table emp(
    eid number primary key,
    ename varchar2(20),
    egender varchar2(10),
    eage number,
    ebirth date
)
prtition by range ( eid ) interval (1000)
(
    partition p1 values less than(1001)
);

insert into emp(eid) values(1100);
insert into emp(eid) values(2800);
select * from user_tab_partitions;
```

【例】 创建员工表，按员工生日分区，第一个分区存生日在1990年1月1日之前的员工，以后每年建一个分区。

```
create table emp(
    eid number primary key,
    ename varchar2(20),
    egender varchar2(10),
    eage number,
    ebirth date
)
prtition by range ( ebirth ) interval ( numtoyminterval (1, 'year') )
(
    partition p1 values less than(to_date('1990 - 01 - 01', 'yyyy - mm - dd' ) )
);
```

```
insert into emp(ebirth) values(to_date('1995-08-01', 'yyyy-mm-dd'));
insert into emp(ebirth) values(to_date('1999-05-08', 'yyyy-mm-dd'));
select * from user_tab_partitions;
```

5) 范围-列表组合分区

语法：

```
CREATE TABLE table_name(
column_name datatype
[, column_name datatype ]
)
PARTITION BY RANGE ( column_name )
SUBPARTITION BY LIST( column_name )
(
    PARTITION partition_name VALUES LESS THAN { literal | MAXVALUE}
    (
        SUBPARTITION partition_name VALUES { literal | DEFAULT}
        [, …]
    )
    [, …]
)
```

属性说明：

(1) table_name

指定要创建的表名称。

(2) column_name datatype

指定表中列的名称及数据类型，如果存在多个列，则用","隔开。

(3) PARTITION BY RANGE

通过关键字 PARTITION BY RANGE 声明是范围分区，括号中 column_name 指定范围分区列。

(4) SUBPARTITION BY LIST

通过关键字 SUBPARTITION BY LIST 声明列表子分区，括号中 column_name 指定列表分区列。

(5) PARTITION

partition_name 指定分区名，literal 指定一个具体值用来限定范围。该语句的作用是定义值小于 literal 的在一个分区中，定义最后一个分区时指定值小于 MAXVALUE。

(6) SUBPARTITION

partition_name 指定分区名，literal 指定一个具体值用来限定范围。该语句的作用是定义值等于 literal 的在一个分区中，定义最后一个分区时指定值为 DEFAULT。

【例】 创建员工表，首先按员工编号分区，第一个分区员工号≤1000号，第二个分区1001～2000号，其余放第三个分区；然后按性别分区，每个分区的第一个子分区存男性员工，第二个分区存女性员工。

```
create table emp(
    eid number primary key,
    ename varchar2(20),
    egender varchar2(10),
    eage number,
    ebirth date
)
prtition by range ( eid )
subpartition by list ( egender )
(
    partition p1 values less than(1001)
    (
        subpartition p1_1 values ('男'),
        subpartition p1_2 values ('女')
    ),
    partition p2 values less than(2001)
    (
        subpartition p2_1 values ('男'),
        subpartition p2_2 values ('女')
    ),
    partition p3 values less than(maxvalue)
);
```

6）范围-范围组合分区

语法：

```
CREATE TABLE table_name(
column_name datatype
[, column_name datatype ]
)
PARTITION BY RANGE ( column_name )
SUBPARTITION BY RANGE( column_name )
(
    PARTITION partition_name VALUES LESS THAN { literal | MAXVALUE}
    (
        SUBPARTITION partition_name VALUES LESS THAN { literal | MAXVALUE}
        [, …]
    )
    [, …]
)
```

【例】 创建员工表，首先按员工编号分区，第一个分区员工号≤1000号，第二个分区1001～2000号，其余放第三个分区；然后按年龄分区，每个分区的第一个子分区存年龄小于30岁员工，第二个分区存年龄介于30～50岁的员工，其余放第三个分区。

```
create table emp(
    eid number primary key,
    ename varchar2(20),
```

```
        egender varchar2(10),
        eage number,
        ebirth date
)
prtition by range ( eid )
subpartition by range( eage )
(
    partition p1 values less than(1001)
    (
        subpartition p1_1 values less than (30 ),
        subpartition p1_1 values less than (50 ),
        subpartition p1_1 values less than (maxvalue )
    ),
    partition p2 values less than(2001)
    (
        subpartition p1_1 values less than (30 ),
        subpartition p1_1 values less than (50 ),
        subpartition p1_1 values less than (maxvalue )
    ),
    partition p3 values less than(maxvalue)
);
```

2. 修改表分区

语法:

```
ALTER TABLE table_name
[ ADD PARTITION partition_description ]
[ DROP PARTITION partition_name ]
[ MERGE PARTITIONS partition_name,[, … ] INTO PARTITION partition_name]
[ SPLIT PARTITION partition_name AT (literal) INTO (PARTITION partition_name [, … ] ) ]
[ TRUNCATE PARTITION partition_name ]
[ RENAME PARTITION old_name TO new_name ];
```

属性说明:

1) table_name

指定要修改的表名称。

2) ADD PARTITION

在表中增加分区,具体描述与分区类型相关,参见创建表分区语法。

3) DROP PARTITION

通过分区名将分区删除。

4) MERGE PARTITIONS

将多个分区数据合并成一个分区,要求合并到范围大的分区。

5) SPLIT PARTITION

将分区按照 literal 值进行拆分,分别存储在 INTO 后指定的分区中。

6) TRUNCATE PARTITION

截断分区数据。

7) RENAME PARTITION

修改分区名字。

【例】 创建范围分区表,进行分区基本操作。

```
第一步 创建分区
SQL> create table emp
(
    eid number,
    ename varchar2(20),
    egender varchar2(10)
)
partition by range(eid)
(
    partition p1 values less than(1000),
    partition p2 values less than(2000),
    partition p3 values less than(maxvalue)
);
Table created.

SQL> insert into emp values(500,'李晓','male');
1 row created.

SQL> insert into emp values(1500,'张良','female');
1 row created.

SQL> select * from emp partition(p2);
      EID   ENAME  EGENDER
    -------- ----- -----------
      1500   张良   female

第二步 合并分区
SQL> alter table emp merge partitions P1,P2 into partition P2;
Table altered.

SQL> select * from emp partition(p2);
      EID   ENAME  EGENDER
    -------- ----- -----------
       500   李晓   male
      1500   张良   female

第三步 拆分分区
SQL> alter table emp
    split partition p2 at(1000) into (partition p1,partition p2);
Table altered.

SQL> select * from emp partition(p2);
      EID   ENAME  EGENDER
    -------- ----- -----------
      1500   张良   female
```

第四步 截断分区
```
SQL> alter table emp truncate partition p2;
Table truncated.

SQL> select * from emp partition(p2);
no rows selected
```

第五步 修改分区名
```
SQL> alter table emp rename partition p3 to pmax;
Table altered.
```

8.1.6 相关视图

1. DBA_TABLES

DBA_TABLES 视图显示数据库中所有用户表信息，参见表 8-1。

表 8-1　DBA_TABLES 主要列介绍

列　名	描　述
OWNER	表所属用户
TABLE_NAME	表名
TABLESPACE_NAME	表所在表空间名
CLUSTER_NAME	所属簇名
LOGGING	日志状态，logging 或 nologging
NUM_ROWS	记录行数
BLOCKS	使用的数据块数
EMPTY_BLOCKS	空数据块数
IOT_TYPE	是否索引表
TEMPORARY	是否临时表
BUFFER_POOL	默认存储缓冲池
READ_ONLY	是否只读

2. DBA_TAB_COLUMNS

DBA_TAB_COLUMNS 显示数据库中所有表、视图的列信息，参见表 8-2。

表 8-2　DBA_TAB_COLUMNS 主要列介绍

列　名	描　述
OWNER	表所属用户
TABLE_NAME	表名
COLUMN_NAME	列名
DATE_TYPE	数据类型
DATE_LENGTH	数据长度
DATE_PRECISION	数据精度

8.2 视 图

8.2.1 普通视图

视图本质上是一个存储的查询语句。查询视图实际上查询的是创建视图语句中的基本表,视图只是基本表的一部分或全部数据的一种展现形式。

1. 创建视图

创建视图语法如下:

```
CREATE OR REPLACE [ NO FORCE | FORCE ] VIEW view_name
[ (column_alias, [,…] ) ]
AS subquery
[ WITH { READ ONLY | CHECK OPTION} ];
```

属性说明:

1) view_name

指定要创建的视图名称。

2) NO FORCE | FORCE

指定当基表不存在时是否强制创建视图,默认是 no force。

3) column_alias,[,…]

指定创建的列名,如果未指定则查询语句的列名就是视图的列名。

4) AS

指定创建视图的查询语句,可以是基于单表查询,也可以是连接查询。

5) WITH { READ ONLY | CHECK OPTION}

with read only 指定视图只读,with check option 指定对视图数据操作时是否进行条件检查。

【例】 创建视图。

```
第一步 创建基本表,插入两条数据
SQL> create table salary(name varchar2(20),salary number(8,2));

SQL> insert into salary values('李阳',6000);
SQL> insert into salary values('赵林',3000);

第二步 创建视图 v1,并通过视图插入数据
SQL> create view v1 as select * from salary;
View created.


SQL> select * from v1;
NAME          SALARY
----------    ----------
```

```
李阳            6000
赵林            3000

SQL> insert into salary values('钱亮',2000);
1 row created.

SQL> select * from v1;
NAME            SALARY
----------      ----------
李阳            6000
赵林            3000
钱亮            2000

SQL> select * from salary;
NAME            SALARY
----------      ----------
李阳            6000
赵林            3000
钱亮            2000
```

第三步 创建只读视图 v2,通过视图插入数据报错
```
SQL> create view v2 as select * from salary with read only;
View created.

SQL> select * from v2;
NAME            SALARY
----------      ----------
李阳            6000
赵林            3000
钱亮            2000

SQL> insert into v2 values('王刚',7000);
insert into v2 values('王刚',7000)
            *
ERROR at line 1:
ORA-42399: cannot perform a DML operation on a read-only view
```

第四步 创建视图 v3,通过视图插入数据,一条满足条件,另一条不满足条件
```
SQL> create view v3 as select * from salary where salary>5000 with check option;
View created.

SQL> select * from v3;
NAME            SALARY
----------      ----------
李阳            6000

SQL> insert into v3 values('熊大',6000);
1 row created.
```

```
SQL> insert into v3 values('熊二',4000);
insert into v3 values('熊二',4000)
                *
ERROR at line 1:
ORA-01402: view WITH CHECK OPTION where-clause violation


SQL> select * from v3;
NAME         SALARY
----------   ----------
李阳          6000
熊大          6000
```

2．删除视图

删除视图语法如下：

DROP VIEW view_name [CASCADE CONSTRAINTS]

属性说明：

1) view_name

指定要删除的视图名称。

2) CASCADE CONSTRAINTS

指定删除视图时，删除所有引用该视图的引用约束。

8.2.2 物化视图

物化视图包含创建视图的语句结果，针对每个物化视图 Oracle 数据库自动创建一个数据库表，这个数据库表叫做容器表。物化视图通常只是作为只读用，如果需要通过物化视图更新基表数据可参考 Oracle 数据库的高级复制内容。

1．创建物化视图

创建物化视图语法如下：

```
CREATE MATERIALIZED VIEW view_name
[ (column_alias, [, …] ) ]
[ REFRESH
    [ FAST | COMPLETE | FORCE ]
    [ ON DEMAND | ON COMMIT | START WITH date NEXT date ]
    WITH { PRIMARY KEY | ROWID }
]
[ { ENABLE | DISABLE } QUERY REWRITE ]
AS subquery;
```

属性说明：

1) view_name

指定要创建的物化视图名称。

2) column_alias, [,…]

指定创建的列名，如果未指定则查询语句的列名就是视图的列名。

3) REFRESH

指定视图刷新方法。

4) FAST | COMPLETE | FORCE

指定是否使用增量刷新，fast 方式指定使用物化视图日志来记录基表变化并根据物化视图日志内容来刷新，complete 指定完全刷新，force 指定如果存在物化视图日志执行 fast 方式，否则执行 complete 方式。

5) ON DEMAND | ON COMMIT | START WITH date NEXT date

指定视图刷新触发机制，on demand 方式使用 Oracle 数据库包 dbms_mview 中的过程来手动完成刷新，refresh on demand 是默认方法，on commit 方式指定基表事务提交时立即刷新，START WITH date NEXT date 方式指定时间间隔定时刷新。

6) WITH { PRIMARY KEY | ROWID }

指定视图刷新数据时依据条件，创建物化视图的基表一般都有主键，with primary key 是默认方式，如果基表没有主键，则依据行号 rowid 来刷新数据，这时必须显式指定 with rowid。

7) { ENABLE | DISABLE } QUERY REWRITE

指定是否支持查询重写特性。

【例】 创建手动刷新的物化视图。

```
SQL > create table t01(id number primary key, name varchar2(20));
Table created.

SQL > create materialized view mv1 as select * from t01;
Materialized view created.

SQL > insert into t01 values(1,'a');
1 row created.

SQL > commit;
Commit complete.
SQL > select * from mv1;
No rows selected.

SQL > begin
    dbms_mview.refresh(list => 'MV1');
   end;
   /
PL/SQL procedure successfully completed.

SQL > select * from mv1;
    ID NAME
------ --------------
     1 a
```

【例】 创建提交刷新的物化视图。

```
SQL> create materialized view mv2 refresh on commit as select * from t01;
Materialized view created.

SQL> insert into t01 values(2,'b');
1 row created.

SQL> commit;
Commit complete.

SQL> select * from mv2;
    ID  NAME
------  ------------
     1  a
     2  b
```

【例】 创建定时刷新（每隔 5min）的物化视图。

```
SQL> create materialized view mv3 refresh
     start with sysdate next sysdate + 5/1440
     as select * from t01;
Materialized view created.

SQL> insert into t01 values(3,'c');
1 row created.

SQL> commit;
Commit complete.

SQL> select * from mv3;
    ID  NAME
------  ------------
     1  a
     2  b

5min 后再查询
SQL> select * from mv3;
    ID  NAME
------  ------------
     1  a
     2  b
     3  c
```

【例】 创建物化视图日志,再创建快速手动刷新视图。

```
SQL > create materialized view log on t01;
Materialized view log created.

SQL > create materialized view mv4 refresh fast on demand as select * from t01;
Materialized view created.

SQL > insert into t01 values(4,'d');
1 row created.

SQL > select * from MLOG$_T01;
    ID  SNAPTIME$$    DMLTYPE$$    OLD_NEW$$    CHANGE_VECTOR$$    XID$$
 ----- ------------ ------------ ------------ ------------------- ---------------
     4 01-1月 -00  I             N             FE                  844489354643779

SQL > begin
    dbms_mview.refresh(list = >'MV4');
    end;
    /

PL/SQL procedure successfully completed.

SQL > select * from mv4;
    ID NAME
--------- --------------------
     1 a
     2 b
     3 c
     4 d

SQL > select * from MLOG$_T01;
No rows selected.
```

2. 删除物化视图

删除物化视图语法如下:

```
DROP MATERIALIZED VIEW view_name [ PRESERVE TABLE];
```

属性说明:

1) view_name

指定要删除的视图名称。

2) PRESERVE TABLE

指定当物化视图删除时,保留物化视图的容器表。

8.2.3 相关视图

1. DBA_VIEWS

DBA_VIEWS 视图显示数据库中所有用户视图信息,参见表 8-3。

表 8-3 DBA_VIEWS 主要列介绍

列　　名	描　　述
OWNER	所属用户
VIEW_NAME	视图名
TEXT	创建视图语句
READ_ONLY	是否只读

2. DBA_MVIEWS

DBA_MVIEWS 视图显示数据库中所有用户物化视图信息,参见表 8-4。

表 8-4 DBA_MVIEWS 主要列介绍

列　　名	描　　述
OWNER	所属用户
MVIEW_NAME	视图名
CONTAINER_NAME	容器表名
QUERY	创建视图语句
UPDATABLE	是否可更新
REWRITE_ENABLED	是否支持查询重写
REFRESH_MODE	刷新模式,on demand 或 on commit
REFRESH_METHOD	刷新方法,fast、complete 或 force

8.3 索　　引

索引是可选的数据结构,通常与表或簇表关联,建立索引可提升数据访问速度,索引对于数据访问的作用与目录对于书的作用是一样的。

8.3.1 索引类别

Oracle 数据库中主要的索引类型是 B 树索引和位图索引。

1. B 树索引

B 树索引(B-trees,balanced trees),是最常用的数据库索引类型。索引数据包含索引键值和对应行的 ROWID,一个索引键值对应一行数据,唯一索引中是一个键值唯一对应一行数据,非唯一索引中一个键值可能对应多行数据,在索引中分多行存储,索引数据是排序的。图 8-1 显示了 B 树索引结构。

2. 位图索引

一个索引条目指向多行,适用于相异基数低的数据,键值可枚举,比如员工的性别、出生月份等。不适合在线事务系统(OLTP),数据经常变,会导致位图数据大量更新,在线分析(OLAP)系统应用较多,表数据基本不变。图 8-2 显示了位图索引结构。

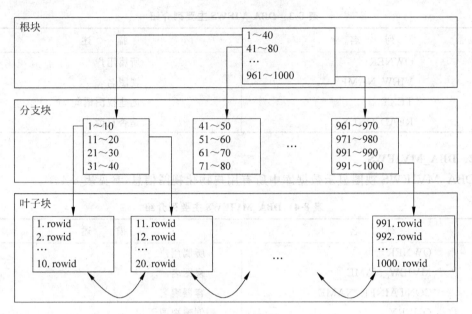

图 8-1　B 树索引

键值/行号	1	2	3	4	5	6	7	…	100
男	1	0	0	1	1	0	0		1
女	0	1	1	0	0	1	1		0

图 8-2　位图索引

除上述两种主要索引类型外，还可细分为单列索引、复合索引、基于函数的索引、唯一索引、非唯一索引、反向索引以及分区索引。

1）单列索引

只在表的一列上创建索引，同一列上不能创建多个单列索引。

2）复合索引

在表的多列上创建索引。

3）基于函数的索引

索引键值基于函数运算的结果产生。

4）唯一索引

每个索引键值唯一对应到一行数据。

5）非唯一索引

每个索引键值可对应一行或多行数据。

6）反向索引

将索引键值字节序颠倒，打散原索引的排序数据，一般用在单个索引块出现访问压力情况下。

7）分区索引

在表分区的情况下，索引可以分区也可以不分区，分区索引分为全局分区索引和局部分区索引，如图 8-3 所示。

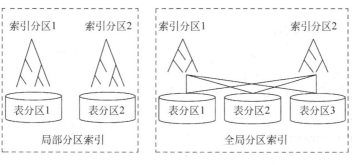

图 8-3 分区索引

8.3.2 索引维护

1. 创建索引

创建索引语法如下：

```
CREATE [ UNIQUE | BITMAP ] INDEX index
ON table (index_expr [ ASC | DESC ])
[ GLOBAL PARTITION BY { RANGE | HASH } ( column ) ( PARTITION description)
 | LOCAL ]
[ TABLESPACE tablespace ]
[ REVERSE ]
[ VISIBLE | INVISIBLE ];
```

属性说明：

1) UNIQUE | BITMAP

指定要创建的索引类型，默认情况是非唯一 B 树索引，unique 指定是唯一 B 树索引，bitmap 指定是位图索引。位图索引通常用于在线分析（OLAP）系统。

2) index

指定要创建的索引名字。

3) ON

指定在表的哪些列上创建索引，index_expr 可以是表的列名，也可以是表达式，如果指定一列就是单列索引，如果指定多列就是复合索引，如果使用函数或其他运算就是基于函数的索引，ASC 指定索引数据升序排序，DESC 指定是降序排序，默认是 ASC。

4) GLOBAL PARTITION | LOCAL

当表已经分区的情况下，相应的索引有三种选择：一是不分区；二是单独分区，与表分区方式不一样；三是本地分区，分区方式和表一致。省略该语句，就是不分区，不管表是否分区，指定 LOCAL 就是本地分区，另外就是通过指定 GLOBAL PARTITION BY 语句进行单独分区。

5) TABLESPACE

指定索引存储表空间。

6) REVERSE

指定创建反向索引，一般在索引块数据访问频繁压力较大时，通过反向创建索引将索引数据分散开。

【例】 在表上创建索引。

```
SQL> create table emp(
      eno number,
      ename varchar2(20),
      eage number,
      egender varchar2(20)
     );
Table created.

SQL> create unique index emp_idx1 on emp(eno);
Index created.

SQL> create index emp_idx2 on emp(eage,egender);
Index created.

SQL> create index emp_idx3 on emp(mod(eage,5));
Index created.
```

2. 修改索引

修改索引语法如下:

```
ALTER INDEX index
[ COALESCE ]
[ REBUILD ]
[ RENAME TO new_name ];
```

属性说明:

1) index

指定要修改的索引名字。

2) COALESCE

当表中数据频繁插入和删除时,对应的索引数据会出现不连续,该语句可以合并索引的叶子块上的数据。

3) REBUILD

当表中数据频繁插入和删除时,对应的索引数据会出现不连续,该语句重新构建整个索引。

4) RENAME TO

修改索引名字。

3. 删除索引

删除索引语法如下:

```
DROP INDEX index;
```

8.3.3 相关视图

1. DBA_INDEXES

DBA_INDEXES 视图显示数据库中所有索引信息,参见表 8-5。

表 8-5 DBA_INDEXES 主要列介绍

列 名	描 述
OWNER	所属用户
INDEX_NAME	索引名
INDEX_TYPE	索引类型,位图索引、索引组织表、函数索引等
TABLE_OWNER	表所属用户
TABLE_NAME	表名
UNIQUENESS	是否唯一索引
TABLESPACE_NAME	表空间名
BLEVEL	深度
LEAF_BLOCKS	叶子块数
DISTINCT_KEYS	不同键值数
CULSTERING_FACTOR	聚集因子

2. DBA_IND_COLUMNS

DBA_IND_COLUMNS 视图显示数据库中所有索引列信息,参见表 8-6。

表 8-6 DBA_IND_COLUMNS 主要列介绍

列 名	描 述
INDEX_OWNER	所属用户
INDEX_NAME	索引名
TABLE_OWNER	表所属用户
TABLE_NAME	表名
COLUMN_NAME	索引列名
COLUMN_POSITION	索引列顺序
DESCEND	该列是否降序排列

8.4 簇

簇可以包含一个或多个表数据,根据共有列将表数据进行组织,当根据共有列查询多表数据时可减少磁盘读写数据量。

1. 创建簇

创建簇语法如下:

```
CREATE CLUSTER cluster ( column datatype )
[ SIZE size ]
[ TABLESPACE tablespace ]
[ INDEX | HASHKEYS integer ];
```

属性说明：

1) cluster

指定要创建的簇名称。

2) column datatype

指定簇键以及数据类型，簇键可以是一个或多个，最多 16 个。

3) SIZE

指定可以存储的和簇键或哈希值相关联的数据大小，最好是一个数据块大小，这样一个数据块上只存放关联数据。

4) TABLESPACE

指定簇存放的表空间。

5) INDEX | HASHKEYS integer

指定簇类型，index 指索引簇，索引簇必须在簇上建索引，index 是默认值。Hashkeys 指定是哈希簇，后面的 integer 指定哈希值数量，创建哈希簇时，Oracle 数据库根据 size 和 integer 乘积预分配空间。

【例】 创建一个索引簇，并在簇上创建表。

```
SQL> create cluster dept_emp(did number) index size 4K;
Cluster created.

SQL> create index dept_emp_idx on cluster dept_emp;
Index created.

SQL> create table dept(
    dno number primary key,
    dname varchar2(20)
    )cluster dept_emp(dno);
Table created.

SQL> create table emp(
    eno number primary key,
    ename varchar2(20),
    dno number references dept(dno)
    )cluster dept_emp(dno);
Table created.
```

2. 删除簇

删除簇语法如下：

```
DROP CLUSTER cluster
[ INCLUDING TABLES [ CASCADE CONSTRAINTS ] ];
```

属性说明：

1) cluster

指定要创建的簇名称。

2) INCLUDING TABLES

指定删除簇上所有的表。

3) CASCADE CONSTRAINTS

指定删除表时,删除所有引用该表的引用约束。

3. 相关视图

DBA_CLUSTERS 视图显示数据库中所有簇信息,参见表 8-7。

表 8-7 DBA_CLUSTERS 主要列介绍

列 名	描 述
OWNER	所属用户
CLUSTER_NAME	簇名
TABLESPACE_NAME	表空间名
CLUSTER_TYPE	簇类型,index 或 hash

8.5 序 列

序列 sequence 是一种数据库对象,用于生成一个正数序列,通常用来填充数字类型的主键列。

1. 创建序列

创建序列语法如下:

```
CREATE SEQUENCE sequence_name
[ INCREMENT BY integer ]
[ START WITH integer ]
[ MAXVALUE integer | NOMAXVALUE ]
[ MINVALUE integer | NOMINVALUE ]
[ CYCLE | NOCYCLE ]
[ CACHE integer | NOCACHE ]
[ ORDER | NOORDER ];
```

属性说明:

1) sequence_name

指定要创建的序列名称。

2) INCREMENT BY

指定序列每次递增值,可为正数或负数,但绝对值要小于最大值和最小值之差,默认为 1。

3) START WITH

指定序列起始值,默认为 1。

4) MAXVALUE integer | NOMAXVALUE

可以通过 maxvalue n 指定序列最大值,如未指定,取默认值 nomaxvalue。升序时 nomaxvalue 为 10^{27},降序时 nomaxvalue 为 -1。maxvalue 指定的值必须大于等于 start with 指定的值,而且必须大于 minvalue 指定的值。

5) MINVALUE integer | NOMINVALUE

可以通过 minvalue n 指定序列最小值，如未指定，取默认值 nominvalue。升序时 nominvalue 为 1，降序时 nominvalue 为 -10^{26}。maxvalue 指定的值必须小于等于 start with 指定的值，而且必须小于 maxvalue 指定的值。

6) CYCLE | NOCYCLE

指定当序列值达到极限后，是否重新开始，默认是 nocycle。

7) CACHE integer | NOCACHE

指定是否为序列预分配值，预分配的好处在于性能有所提高，但一旦数据库异常关闭会导致预分配的值丢失，导致序列值不连续。数据库默认预分配 20 个。如果需要序列必须连续，那么使用 nocache。

8) ORDER | NOORDER

用于指定是否按照请求次序生成整数，数据库默认为 noorder。

【例】 创建一个序列，起始 10，每次递增 2，没有最大值限制。

```
SQL> create sequence seq start with 10 increment by 2;
Sequence created.
```

2. 使用序列

一个序列中包含两个伪列，分别是 currval 和 nextval，可以分别用来取该序列的当前值和下一个值。

currval：返回序列当前值，使用这个伪列时必须先通过检索序列的下一个值对序列进行初始化。

nextval：返回当前序列值增加一个步长后的值。

【例】 创建一个序列，起始 10，每次递增 2，没有最大值限制。

```
SQL> select seq.currval from dual;
select seq.currval from dual
       *
ERROR at line 1:
ORA-08002: sequence SEQ.CURRVAL is not yet defined in this session

SQL> select seq.nextval from dual;
NEXTVAL
----------
10

SQL> select seq.currval from dual;
CURRVAL
----------
10
```

3. 删除序列

删除序列语法如下：

```
DROP SEQUENCE sequence_name;
```

4. 相关视图

DBA_SEQUENCES 视图显示数据库中所有簇信息，参见表 8-8。

表 8-8 DBA_SEQUENCES 主要列介绍

列　　名	描　　述
SEQUENCE_OWNER	所属用户
SEQUENCE_NAME	序列名
MIN_VALUE	最小值
MAX_VALUE	最大值
INCREMENT_BY	递增值
CYCLE_FLAG	是否循环
ORDER_FLAG	是否按要求次序分配
CACHE_SIZE	缓存数量
LAST_NUMBER	最后一次写磁盘的数

8.6　同　义　词

同义词是数据库对象的一个别名，经常用于简化对象访问和提高对象访问的安全性。在使用同义词时，Oracle 数据库将它翻译成对应方案对象的名字。与视图类似，同义词并不占用实际存储空间，只有在数据字典中保存了同义词的定义。在 Oracle 数据库中的大部分数据库对象，如表、视图、同义词、序列、存储过程、函数、Java 类、包、数据类型或其他同义词等，数据库管理员都可以根据实际情况为它们定义同义词。

1. 创建同义词

创建同义词语法如下：

```
CREATE OR REPLACE [ PUBLIC ] SYNONYM synonym FOR object_name[@dblink];
```

属性说明：

1) PUBLIC

指定要创建公有同义词，所有用户都可以访问。

2) synonym

指定要创建的同义词名称。

3) object_name[@dblink]

指定要创建同义词的对象名，如果是远程数据库对象，通过 @dblink 指定远程数据库信息。

【例】 基于表创建一个同义词。

```
SQL> create synonym s1 for t01;
Synonyn created.

SQL> select * from s1;
    ID NAME
------ --------------------
     1 a
     2 b
     3 c
```

2. 删除同义词

删除同义词语法如下：

```
DROP [PUBLIC] SYNONYM synonym;
```

3. 相关视图

DBA_SYNONYMS 视图显示数据库中所有簇信息，参见表 8-9。

表 8-9 DBA_SYNONYMS 主要列介绍

列 名	描 述
OWNER	所属用户
SYNONYM_NAME	同义词名
TABLE_OWNER	建同义词的数据库对象所属用户
TABLE_NAME	建同义词的数据库对象名
DB_LINK	数据库链接名

8.7 数据库链接

数据库链接（DATABASE LINK）是在分布式环境下，为了访问远程数据库而创建的数据通信链路。数据库链接隐藏了对远程数据库访问的复杂性。通常，正在登录的数据库称为本地数据库，另外的一个数据库称为远程数据库。有了数据库链接，可以直接通过数据库链接来访问远程数据库的表。

1. 创建数据库链接

创建数据库链接语法如下：

```
CREATE [PUBLIC] DATABASE LINK dblink
[ CONNECT TO user IDENTIFIED BY password]
USING connect_string;
```

属性说明：

1) PUBLIC

指定要创建公有数据库链接，所有用户都可以访问。

2) dblink

指定要创建的数据库链接名称。

3) CONNECT TO

指定通过指定用户 user 和口令 password 连接远程数据库。如果没有该语句,则数据库使用当前连接数据库的用户名和口令连接远程数据库。

4) USING

指定连接远程数据库的连接串,connect_string 是服务命名。

【例】 假设远程数据库(网络服务命名 remote 指定)用户 u01/u01 拥有表 tbl01,创建数据库链接并访问远程表。

```
SQL> create database link link1
    connect to u01 identified by u01 using 'remote';
SQL> select * from tbl01@link1;
    ID
------
   100
SQL> create synonym tbl01 for  tbl01@link1;
SQL> select * from tbl01;
    ID
------
   100
```

2. 删除数据库链接

删除数据库链接语法如下:

DROP [PUBLIC] DATABASE LINK dblink;

3. 相关视图

DBA_DB_LINKS 视图显示数据库中所有数据库链接信息,参见表 8-10。

表 8-10 DBA_DB_LINKS 主要列介绍

列 名	描 述
OWNER	所属用户
DB_LINK	数据库链接名
USERNAME	连接远程库的用户名
HOST	网络服务命名
CREATED	创建时间

8.8 约 束

8.8.1 约束分类

数据库表的约束是防止无效或有问题的数据输入到表中的一种手段,约束是强加在表上的规则或条件。Oracle 数据库中约束包括主键约束、外键约束、检查约束、唯一约束和非空约束,见表 8-11。

表 8-11　约束种类

约束	类型	定义
PRIMARY KEY	P	主键约束,由一列或多列构成,唯一标识表中的行
UNIQUE	U	指定一列或一组列不能取重复值
NOT NULL	C	规定某一列不允许取空值
CHECK	C	规定一列或一组列的值必须满足指定的约束条件 check(col in(1,2,3)) check(col='123') check(col between 18 and 30)
FOREIGN KEY	R	指定表的外键,可指定删除主表数据的级联方式 ON DELETE {CASCADE|SET NULL}

8.8.2　约束操作

1. 创建约束

创建表时,可在单列或多列上指定约束,如果是在单列上创建约束,可在列级或表级指定约束,如果是在多列上创建约束,则只能在表级指定。

语法如下:

```
CREATE TABLE table_name(
    column_name datatype [CONSTRAINT constraint_name]约束声明
    [,column_name datatype [CONSTRAINT constraint_name]约束声明]
    [,[CONSTRAINT constraint_name]约束声明]
);
```

约束定义时,可以通过关键字 CONSTRAINT 指定约束名称,否则由系统自动产生。

1) 创建主键约束

语法如下:

```
CREATE TABLE table_name(
    column_name datatype [CONSTRAINT constraint_name] PRIMARY KEY
    [,column_name datatype]
);
```

或

```
CREATE TABLE table_name(
    column_name datatype ,
    [,column_name datatype]
    [CONSTRAINT constraint_name] PRIMARY KEY(column_name [,…] )
);
```

【例】　创建学生表,学号作为主键。

```
create table stu(
    stuno number CONSTRAINT stu_pk PRIMARY KEY,
    name varchar2(20)
);
```

或

```
create table stu(
    stuno number,
    name varchar2(20),
    CONSTRAINT stu_pk PRIMARY KEY(id)
);
```

2）创建唯一约束

语法如下：

```
CREATE TABLE table_name(
    column_name datatype [CONSTRAINT constraint_name] UNIQUE
    [,column_name datatype]
);
```

或

```
CREATE TABLE table_name(
    column_name datatype ,
    [,column_name datatype]
    [CONSTRAINT constraint_name] UNIQUE(column_name [, … ] )
);
```

【例】 创建学生表，学号为主键，身份证号唯一。

```
create table stu(
    stuno number constraint stu_pk primary key,
    iden varchar2(18) CONSTRAINT stu_iden_uni UNIQUE,
    name varchar2(20)
);
```

或

```
create table stu(
    stuno number,
    iden varchar2(18),
    name varchar2(20),
    CONSTRAINT stu_iden_uni UNIQUE(iden)
);
```

3）创建非空约束

语法如下：

```
CREATE TABLE table_name(
    column_name datatype [CONSTRAINT constraint_name] NOT NULL
    [,column_name datatype]
);
```

【例】 创建学生表，学号为主键，身份证号非空。

```
create table stu(
    stuno number constraint stu_pk primary key,
    iden varchar2(18) CONSTRAINT stu_iden_nn NOT NULL,
    name varchar2(20)
);
```

4）创建检查约束

语法如下：

```
CREATE TABLE table_name(
    column_name datatype [CONSTRAINT constraint_name] CHECK(condition)
    [,column_name datatype]
);
```

或

```
CREATE TABLE table_name(
    column_name datatype ,
    [,column_name datatype]
    [CONSTRAINT constraint_name] CHECK(condition)
);
```

【例】 创建学生表，学号为主键，年龄介于18～22岁，性别是男或女。

```
create table stu(
    stuno number constraint stu_pk primary key,
    name varchar2(20),
    age number CONSTRAINT stu_age_ck CHECK( age between 18 and 22),
    gender varchar2(2) CONSTRAINT stu_gender_ck CHECK( gender = '男' or gender = '女')
);
```

或

```
create table stu(
    stuno number constraint stu_pk primary key,
    name varchar2(20),
    age number,
    gender varchar2(2),
    CONSTRAINT stu_age_ck CHECK( age >= 18 and age <= 20),
    CONSTRAINT stu_gender_ck CHECK( gender in ('男', '女'))
);
```

5）创建外键约束

语法如下：

```
CREATE TABLE table_name(
    column_name datatype [CONSTRAINT constraint_name]    REFERENCES
            table_name(column_name) [ON DELETE CASCADE| SET NULL]
    [,column_name datatype]
);
```

或

```
CREATE TABLE table_name(
    column_name datatype ,
    [,column_name datatype]
    [CONSTRAINT constraint_name] FOREIGN KEY(column_name)
      REFERENCES table_name(column_name) [ON DELETE CASCADE| SET NULL]
);
```

其中，REFERENCES 语句指定该外键引用的表和列，ON DELETE CASCADE 指当主表数据删除时从表数据发生级联删除，ON DELETE SET NULL 指当主表数据删除时从表数据相应外键列值置 NULL。

【例】 创建部门表，部门编号为主键；创建员工表，员工号为主键，部门号为外键，引用部门表的部门编号列，且当部门表部门数据删除时，员工表部门编号列置 NULL。

```
create table dept(
    dno number constraint dept_pk primary key,
    name varchar2(20)
);

create table emp(
    eid number constraint emp_pk primary key,
    ename varchar2(20),
    did number CONSTRAINT emp_did_fk REFERENCES dept(dno) ON DELETE SET NULL
);
```

或

```
create table emp(
    eid number constraint emp_pk primary key,
    ename varchar2(20),
    did number,
    CONSTRAINT emp_did_fk FOREIGN KEY(did)   REFERENCES dept(dno) ON DELETE SET NULL
);
```

2. 添加约束

语法如下：

```
ALTER TABLE table_name ADD [CONSTRAINT constraint_name] 约束声明;
```

例:

```
alter table emp add primary key(eid);
alter table emp add unique(name);
alter table emp add foreign key(did) references dept(dno);
```

3. 删除约束

语法如下:

```
ALTER TABLE table_name DROP CONSTRAINT constraint_name
    [CASCADE] [KEEP INDEX]
```

如果删除约束同时删除所有引用它的其他约束,指定 CASCADE 参数,对于主键和唯一约束删除时默认删除索引,如需保留索引可使用 KEEP INDEX。

例:

```
alter table emp drop constraint emp_pk;
alter table dept drop constraint dept_pk cascade;
```

4. 设置约束状态

语法:

```
ALTER TABLE table_name MODIRY CONSTRAINT constraint_name
    {ENABLE [NOVALIDATE] | DISABLE[VALIDATE]}
```

ENABLE 状态启用约束并校验表中已有数据和新插入数据,ENABLE NOVALIDATE 启用约束但不校验表中已有数据。DISABLE 禁用约束,DISABLE VALIDATE 禁用约束同时禁止表中的任何 DML 操作,但可使其他工具导入导出数据。

例:

```
alter table emp modify constraint emp_fk DISABLE;
alter table emp modify constraint emp_fk ENABLE NOVALIDATE;
```

8.8.3 相关视图

1. DBA_CONSTRAINTS

DBA_CONSTRAINTS 视图显示数据库中所有约束信息,参见表 8-12。

表 8-12 DBA_CONSTRAINTS 主要列介绍

列 名	描 述
OWNER	所属用户
TABLE_NAME	表名
CONSTRAINT_NAME	约束名
CONSTRAINT_TYPE	约束类型
SEARCH_CONDITION	约束条件
R_OWNER	外键引用的约束所属用户
R_CONSTRAINT_NAME	空外键引用的约束名称
DELETE_RULE	从表数据删除规则
STATUS	约束状态

2. DBA_CONS_COLUMNS

DBA_CONS_COLUMNS 视图显示数据库中所有约束列信息，参见表 8-13。

表 8-13 DBA_CONS_COLUMNS 主要列介绍

列 名	描 述
OWNER	所属用户
TABLE_NAME	表名
CONSTRAINT_NAME	约束名
COLUMN_NAME	约束列
POSITION	约束列顺序

小 结

本章讲解了 Oracle 数据库中常见的数据库对象，包括表、视图、索引、簇、序列、同义词、数据库链接以及表的约束，并重点讲解了其中的表、视图和索引。表这部分首先介绍了常用数据类型，然后讲解了常见的堆表、临时表、索引组织表和分区表，并分别介绍了应用场景；视图这部分除了讲解普通视图的使用外，又讲解了物化视图的使用；索引这部分首先讲解了索引的作用，然后重点讲解了 B 树索引和位图索引，同时介绍了索引的其他概念，比如单列索引、复合索引、基于函数索引、唯一索引、非唯一索引以及分区索引等。

思 考 题

1. 创建一个表 dept，包括 id 和 name 两列，存储在 tbs 表空间上，初始分区空间 10MB。
2. 创建一个表 emp，包括 id、name、gender 和 age 4 列，按 gender 分区，第一个分区存储在 tbs1 表空间上，第二个分区存储在 tbs2 表空间上。
3. 基于表 emp 创建一个视图 empv1，显示所有年龄大于 30 岁的员工，带检查约束。
4. 基于表 emp 创建一个物化视图 empmv1，显示所有年龄大于 50 岁的员工信息，视图数据随 emp 表提交而更新。
5. 在表 emp 的 age 和 gender 列上创建一个复合索引 idx1。
6. 在表 dept 的 id 列上增加主键约束，在表 emp 上增加列 did、外键、引用 dept 表的 id 列。
7. 创建序列 seq1，起始值 10，每次递增 5，最大 10 000，循环。
8. 为表 emp 创建同义词 emps。

第 9 章　SQL 开发

SQL 是访问数据库的标准语言,熟练掌握 SQL 语法并理解 SQL 运行原理是开发高效数据库应用程序的基础。本章重点讲解数据操纵语句(DML)和查询语句(SELECT)的使用。

9.1　结构化查询语言(SQL)简介

SQL 是访问关系数据库的标准语言,语法简单易用。SQL 语句分为如下 5 类。

1. 数据定义(Data Definition Language,DDL)语句

用于定义构成数据库的结构,比如表空间、表、索引等。常用 DDL 语句包括以下三类。

(1) CREATE:创建数据结构。比如:CREATE TABLE 用于创建表,CREATE TABLESPACE 用于创建表空间,CREATE INDEX 用于创建索引。

(2) ALTER:修改数据结构。比如:ALTER TABLE 用于修改表的结构,可以增加列、修改列或删除列。

(3) DROP:删除数据结构。比如:DROP TABLE 用于删除表。

2. 数据操纵(Data Manipulation Language,DML)语句

用于修改表中的数据,可增加数据、删除数据或修改数据。DML 语句包括以下语句。

(1) INSERT:向表中插入新数据。

(2) DELETE:删除表中数据。

(3) UPDATE:修改表中数据。

3. 查询语句(SELECT)

用于查询数据库表中的数据,可以查询单个表的数据,也可以查询多个表的数据。

4. 事务控制(Transaction Control Language,TCL)语句

用于控制事务的提交或撤销。TC 语句包括以下语句。

(1) COMMIT:提交事务,永久保存所做的修改。

(2) ROLLBACK:回滚事务,撤销所做的修改。

(3) SAVEPOINT:设置保存点,可用于控制事务回滚的位置。

5. 数据控制(Data Control Language,DCL)语句

用于控制数据库的权限。DCL 语句包括以下语句。

(1) GRANT:授予用户或角色权限。

(2) REVOKE:回收用户或角色权限。

在如上 5 种 SQL 分类中,数据定义语句(DDL)在第 4 章表空间部分和第 8 章 Schema

对象中有大量介绍,数据控制语句(DCL)在第 5 章用户权限部分有介绍,本章主要针对数据操纵语句(DML)、查询语句(SELECT)和事务控制语句(TCL)进行详细讲解。

9.2 数据操纵语句

数据操纵语句(DML)主要包括插入语句 INSERT、修改数据语句 UPDATE 和删除数据语句 DELETE,另外还有一个表数据合并语句 MERGE,MERGE 语句主要由 INSERT 语句和 UPDATE 语句完成合并功能。

9.2.1 INSERT 语句

使用 INSERT 语句可完成向表中插入数据,可以执行单表插入,也可以执行多表插入,插入的数据可以是单条也可以是多条。

1. 单表插入

语法:

```
INSERT INTO table_name [t_alias] [(column_list)]
{ VALUES ( value1,… )
 | subquery }
```

属性说明:

1) table_name

指定要插入数据的表名称。

2) t_alias

指定表的别名。

3) column_list

指定插入数据列的列表,不写的情况下按表定义时列的顺序插入数据,提供的值必须与列匹配。

4) VALUES

指定插入表的行数据,数据值间以","隔开,与插入数据列的数量和类型匹配。

5) subquery

指定子查询,将子查询的结果全部插入到表中,可实现多条插入。

【例】 插入数据。

```
SQL> create table t01(id number,name varchar2(20),birth date);

插入所有列数据,与值顺序与表定义列的顺序一致
SQL> insert into t01
    values(101,'李国刚',to_date('1991-10-01','yyyy-mm-dd'));

插入指定列的数据,值与指定的列顺序一致
SQL> insert into t01(name,id) values('王洛林',102);
```

使用子查询插入多条数据
```
SQL> create table t02(id number,name varchar2(20));
SQL> insert into t02 select id,name from t01;
SQL> select * from t01;
  ID NAME                 BIRTH
  -- -------------------- ----------
  101 李国刚              1991-10-01
  102 王洛林

SQL> select * from t02;
  ID NAME
  -- --------------------
  101 李国刚
  102 王洛林
```

2. 多表插入

使用 insert 语句和 select 语句实现同时向多个表插入记录，具体可细分为如下三种。

1) 无条件 INSERT ALL

语法：

```
INSERT ALL
insert_into_clause values_clause
[insert_into_clause values_clause]
…
subquery;
```

对于子查询返回的数据行，Oracle 服务器执行每一个 insert_into_clause 一次。

属性说明：

(1) insert_into_clause

指定插入数据的表名和列。

(2) values_clause

指定插入的值。

(3) subquery

指定执行多表插入的数据源，是一个子查询语句。

【例】 插入数据。

```
SQL> create table objects
    as select object_type,object_name,created from user_objects;
SQL> create table object1(objtype varchar2(100),objname varchar2(100));
SQL> create table object2(objname varchar2(100),created date);

SQL> insert all
    into object1 values(object_type,object_name)
    into object2 values(object_name,created)
    select * from objects;
```

【例】 利用 insert all 实现行列转换(insert all 的旋转功能)。

```
假设存在表 week,包含多列,记录一周中每一天的温度
SQL> create table week(
    weekid number,
    monday number(5,2),
    tuesday number(5,2),
    wednesday number(5,2),
    thursday number(5,2),
    friday number(5,2),
    saturday number(5,2),
    sunday number(5,2)
    );
SQL> insert into week values(1,21,20.2,20,18,19.5,19,17.8);

创建 weekday 表,每行记录一天的温度
SQL> create table weekday(
    weekid number,
    weekday varchar2(10),
    degree number(5,2)
    );
SQL> insert all
    into weekday values(weekid, 'Monday ',monday)
    into weekday values(weekid, 'tuesday ',tuesday)
    into weekday values(weekid, 'wednesday ',wednesday)
    into weekday values(weekid, 'thursday ',thursday)
    into weekday values(weekid, 'friday ',friday)
    into weekday values(weekid, 'saturday ',saturday)
    into weekday values(weekid, 'sunday ',sunday)
    select * from week;

SQL> select * from weekday;
    WEEKID  WEEKDAY         DEGREE
    ------  ---------       ------
         1  Monday              21
         1  tuesday           20.2
         1  wednesday           20
         1  thursday            18
         1  friday            19.5
         1  saturday            19
         1  sunday            17.8
```

2) 条件 INSERT ALL

语法:

```
INSERT ALL
WHEN condition THEN insert_into_clause values_clause
[WHEN condition THEN insert_into_clause values_clause]
...
[ELSE insert_into_clause values_clause]
```

subquery;

对于子查询返回的数据行,Oracle 服务器判断每个 when 语句指定的条件,条件满足执行 insert_into_clause。

属性说明:

(1) WHEN condition THEN

通过 condition 指定过滤条件,条件满足执行 THEN 后面的插入语句。

(2) ELSE

WHEN 语句条件都不成立时执行该语句后的插入语句。

【例】 插入数据。

```
SQL> create table salary(id number,name varchar2(20),salary number);
SQL> insert into salary values(101,'刘芳菲',3000);
SQL> insert into salary values(102,'郑明明',5000);
SQL> create table salary1(id number,name varchar2(20),salary number);
SQL> create table salary2(id number,name varchar2(20),salary number);

SQL> insert all
    when salary>2000 then into salary1 values(id,name,salary)
    when salary>4000 then into salary2 values(id,name,salary)
    select * from salary;

SQL> select * from salary1;
        ID NAME                 SALARY
---------- -------------------- --------
       101 刘芳菲                  3000
       102 郑明明                  5000

SQL> select * from salary2;
        ID NAME                 SALARY
---------- -------------------- --------
       102 郑明明                  5000
```

3) 条件 INSERT FIRST

语法:

```
INSERT FIRST
WHEN condition THEN insert_into_clause values_clause
[WHEN condition THEN insert_into_clause values_clause]
...
[ELSE] [insert_into_clause values_clause]
subquery;
```

对于子查询返回的数据行,Oracle 服务器判断第一个 when 语句指定的条件,条件满足执行 insert_into_clause,然后跳过后面的 when 子句,如果条件不满足,再判断下一个 when 子句。

属性说明:

(1) WHEN condition THEN

通过 condition 指定过滤条件，条件满足执行 THEN 后面的插入语句。

(2) ELSE

WHEN 语句条件都不成立时执行该语句后的插入语句。

【例】 插入数据。

```
SQL> create table salary(id number,name varchar2(20),salary number);
SQL> insert into salary values(101,'刘芳菲',3000);
SQL> insert into salary values(102,'郑明明',5000);
SQL> create table salary1(id number,name varchar2(20),salary number);
SQL> create table salary2(id number,name varchar2(20),salary number);

SQL> insert first
    when salary>4000 then into salary2 values(id,name,salary)
    when salary>2000 then into salary1 values(id,name,salary)
    select * from salary;

SQL> select * from salary1;
        ID  NAME                 SALARY
---------- -------------------- --------
       101  刘芳菲                  3000

SQL> select * from salary2;
        ID  NAME                 SALARY
---------- -------------------- --------
       102  郑明明                  5000
```

9.2.2 UPDATE 语句

语法：

```
UPDATE table_name [t_alias]
SET column = {expr | subquery} [, …]
WHERE condition;
```

属性说明：

(1) table_name

指定要修改数据的表名称。

(2) t_alias

指定表的别名。

(3) SET column={expr | subquery}

指定要修改的列，列的值可以是表达式 expr 或子查询结果 subquery，修改多个列，中间用","隔开。

(4) WHERE

指定要修改数据需满足的条件 condition。

【例】 修改数据。

```
SQL> create table salary(id number,name varchar2(20),salary number);
SQL> insert into salary values(101,'刘芳菲',3000);
SQL> insert into salary values(102,'郑明明',5000);
SQL> select * from salary;
        ID NAME                  SALARY
---------- ---------------- ----------
       101 刘芳菲                  3000
       102 郑明明                  5000

SQL> update salary
    set name = '李小萌',salary = (select avg(salary) from salary)
    where id = 101;
SQL> select * from salary;
        ID NAME                  SALARY
---------- ---------------- ----------
       101 李小萌                  4000
       102 郑明明                  5000
```

9.2.3 DELETE 语句

语法：

```
DELETE [FROM] table_name
[WHERE condition];
```

属性说明：

（1）table_name

指定要删除数据的表名称。

（2）t_alias

指定表的别名。

（3）WHERE

指定要删除数据需满足的条件 condition，如果没有指定该语句，则删除表中所有数据。

【例】 删除数据。

```
SQL> create table salary(id number,name varchar2(20),salary number);
SQL> insert into salary values(101,'刘芳菲',3000);
SQL> insert into salary values(102,'郑明明',5000);
SQL> select * from salary;
        ID NAME                  SALARY
---------- ---------------- ----------
       101 刘芳菲                  3000
       102 郑明明                  5000

SQL> delete from salary;
SQL> select * from salary;
no rows selected.
```

9.2.4 MERGE 语句

使用 MERGE 语句可将一个表的数据合并到另一个表,语法如下:

```
MERGE INTO target_table_name [t_alias]
USING source_table_name [t_alias]
ON (condition)
WHEN MATCHED THEN
    update_statement
WHEN NOT MATCHED THEN
    insert_statement;
```

属性说明:

(1) target_table_name

指定要合并数据的目标表的名称。

(2) t_alias

指定表的别名。

(3) USING source_table_name

指定要被合并表的名称。

(4) ON(condition)

指定表数据的关联条件。

(5) WHEN MATCHED THEN

当发现两个表的数据符合关联条件时,执行后面的 update_statement 修改目标表的数据。

(6) WHEN NOT MATCHED THEN

当发现两个表的数据不匹配时,执行后面的 insert_statement 语句将数据插入目标表。

【例】 合并数据。

```
假设存在学生信息表 stu1 和 stu2,表结构相同,现将表 stu2 合并到 stu1,如果 stu1 中存在学号相
同的学生用 stu2 中的信息替换,如 stu1 中不存在而 stu2 中存在则插入 stu1 中
SQL> create table stu1(stuid number,name varchar2(20),age number);
SQL> insert into stu1 values(101,'张晓明',20);

SQL> create table stu2(stuid number,name varchar2(20),age number);
SQL> insert into stu2 values(101,'张小茗',21);
SQL> insert into stu2 values(102,'李飞',25);

SQL> merge into stu1 s1
    using stu2 s2 on(s1.stuid = s2.stuid)
    when matched then
        update set s1.name = s2.name,s1.age = s2.age
    when not matched then
        insert(s1.stuid,s1.name,s1.age) values(s2.stuid,s2.name,s2.age);
SQL> select * from stu1;
     STUID   NAME                          AGE
     -----   ----------------              -----
     101     张小茗                         21
     102     李飞                           25
```

9.3 查询语句

9.3.1 单表查询

SELECT 语句用于从数据库表中查询数据,最基本的应用是从单个表中查询数据,查询的结果称为结果集(Result Set)。

查询语句基本语法如下:

```
SELECT [DISTINCT ] column_list
FROM table_name
WHERE condition
GROUP BY column_name[,…]
HAVING condition
ORDER BY column_name {ASC|DESC} [,…]
```

属性说明:

(1) SELECT

SELECT 子句后指定查询列表,查询列表可以是星号(*),代表查询所有列,也可以是列名的列表,如 name,age,或者是运算表达式。DISTINCT 指查询不同值。

(2) FROM

FROM 子句后指定要查询的表,也可以是视图名、同义词名,或者是一个子查询语句。

(3) WHERE

WHERE 子句指定数据过滤条件,只返回符合条件的数据。比如只查询年龄大于 20 的学生:select * from tblstudent where age>20。

(4) GROUP BY

GROUP BY 子句指定数据分组统计的分组列,可以是一列,也可以是多列。

(5) HAVING

HAVING 子句指定分组后的数据的过滤条件,只返回符合条件的数据。比如只查询学生数大于 30 的班级编号:select classid from tblstudent group by classid having count(*)>30。

(6) ORDER BY

ORDER BY 子句指定查询结果排序列,数据将按指定列进行排序然后返回,默认是升序排序 ASC,也可以指定降序排序 DESC。

(1) 查询表中列数据

如果希望查询表中所有列的数据,可以指定所有列名或用星号(*)取代列名,使用星号查询时返回列数据的顺序与建表时列顺序相同,下面的两个查询是等价的。

```
SQL> select * from tblstudent;
STUDENTID  STUDENTNAME  STUAGE  CLASSID
---------  -----------  ------  -------
2009001    黎明             20    101
2009002    赵亮             19    101
2009003    李晓             20    102

SQL> select studentid,studentname,stuage,classid from tblstudent;
STUDENTID  STUDENTNAME  STUAGE  CLASSID
---------  -----------  ------  -------
2009001    黎明             20    101
2009002    赵亮             19    101
2009003    李晓             20    102
```

除了查询所有列外,可以根据需求查询指定列,比如:

```
SQL> select studentname,stuage from tblstudent;
STUDENTNAME  STUAGE
-----------  ------
黎明             20
赵亮             19
李晓             20
```

查询时除指定表中列外,也可以使用表达式,比如:

```
SQL> select studentname,stuage + 10 from tblstudent;
STUDENTNAME  STUAGE + 10
-----------  -----------
黎明             30
赵亮             29
李晓             30
```

(2) 指定列的别名

当从一个表中查询数据时,Oracle 将列名转换成大写作为列的标题,如果是表达式,则整个表达式就作为列的标题,比如上面查询中的表达式 stuage+10。这时可以使用列的别名来指定标题,别名指定方法为:

列名或表达式 [as] 别名

别名如果要求按照指定大小写显示或别名中包含空格等字符,需要用双引号将别名引起来。

```
SQL> select studentname name, stuage + 10 "Age" from tblstudent;
NAME                 Age
-------------------  ---
黎明                  30
赵亮                  29
李晓                  30
```

(3) 禁止显示重复的数据

查询学生表中的班级编号,结果如下:

```
SQL> select classid from tblstudent;
CLASSID
-------
    101
    101
    102
```

结果中班级编号 101 出现两次,如果希望显示不重复的数据,可以使用 DISTINCT 或 UNIQUE,下面两个查询结果一样。

```
SQL> select distinct classid from tblstudent;
CLASSID
-------
    101
    102

SQL> select unique classid from tblstudent;
CLASSID
-------
    101
    102
```

(4) ROWID 与 ROWNUM

Oracle 数据库表中的每行数据库都有一个唯一的行号,称为 ROWID,Oracle 内部使用 ROWID 来直接定位数据行,下面的查询除了显示表数据外,同时显示行号。

```
SQL> select rowid, classid, classname from tblclass;
ROWID                CLASSID  CLASSNAME
------------------   -------  ------------------
AAADS7AAFAAAACHAAA       101  软件1班
AAADS7AAFAAAACHAAB       102  软件2班
```

另外一个行号称为 ROWNUM,是指返回结果集中的数据行的编号,下面的查询显示了返回结果的行号。

```
SQL> select rownum, classid, classname from tblclass;
ROWNUM  CLASSID  CLASSNAME
------  -------  ------------------
     1      101  软件1班
     2      102  软件2班
```

可以使用 ROWNUM 控制显示返回结果集中的前多少行,比如显示学生表数据的前两行。

```
SQL> select rownum, studentname from tblstudent where rownum <= 2;
ROWNUM    STUDENTNAME
------    --------------------
     1    黎明
     2    赵亮
```

(5) 使用连接操作符输出结果

使用连接操作符"||"可将输出结果进行连接。

```
SQL> select '学号：' || studentid || ',' || '姓名：' || studentname as 学生信息 from
tblstudent;
学生信息
---------------------------------------------
学号：2009001,姓名：黎明
学号：2009002,姓名：赵亮
学号：2009003,姓名：李晓
学号：2009004,姓名：李晓芳
学号：2009005,姓名：
```

上面的查询将列 studentid 和 studentname 用连接操作符与字符串连接,输出结果变成"学号：×××××,姓名：×××××"的样式。

(6) 指定过滤条件

往往一个数据库表中的数据有很多行,而我们关心的只是其中一部分数据,这时可以指定 WHERE 语句来指定条件,将不符合条件的数据过滤掉,比如只想查询年龄小于 20 的学生：

```
SQL> select * from tblstudent where stuage < 20;

STUDENTID STUDENTNAME STUAGE CLASSID
--------- ----------- ------ -------
2009002   赵亮            19    101
```

WHERE 语句后的条件表达式中常用运算符见表 9-1。

表 9-1 运算符

运算符分类	运 算 符	说 明
关系运算符	＞	大于
	＜	小于
	＝	等于
	＞＝	大于等于
	＜＝	小于等于
	＜＞或!＝	不等于
	＞all，＞＝all，＜all，＜＝all	与一个列表中的所有值比较
	＞any，＞＝any，＜any，＜＝any	与一个列表中的任意值比较

续表

运算符分类	运算符	说明
范围运算符	between … and …	在范围内,比如判断年龄介于 20 和 30 之间,age between 20 and 30
	not between … and …	不在范围内,比如判断年龄小于 20 或大于 30,age not between 20 and 30
集合运算	in	在集合内,比如判断年龄是 22、25、28,age in (22, 25,28)
	not in	不在集合内,比如判断年龄不是 22、25、28,age not in (22,25,28)
模式匹配	like	字符匹配,"%"代表任意个字符,"_"代表单个字符, 比如判断所有姓张的人,name like '张%',判断姓名是三个字且中间字是"晓",name like '_晓_'
	not like	不匹配,比如判断所有不姓张的人,name not like '张%'
空值比较	is null	空,比如判断姓名为空,name is null
	is not null	不空,比如判断姓名不为空,name is not null
逻辑运算	and	与,比如判断年龄是 20 且姓张,age = 20 and name like '张%'
	or	或,比如判断年龄是 20 或 30,age=20 or age=30
	not	非,比如 not in、not between … and …

【例】 查询所有学生信息。

```
SQL> select * from tblstudent;
STUDENTID  STUDENTNAME              STUAGE    CLASSID
---------  --------------------     ------    ------
2009001    黎明                         20       101
2009002    赵亮                         19       101
2009003    李晓                         20       102
2009004    李晓芳                                102
2009005                                 22       101
```

【例】 查询姓李的学生信息。

```
SQL> select * from tblstudent where studentname like '李%';
STUDENTID  STUDENTNAME              STUAGE    CLASSID
---------  --------------------     ------    ------
2009003    李晓                         20       102
2009004    李晓芳                                102
```

【例】 查询年龄为 19 和 20 的学生信息。

```
SQL> select * from tblstudent where stuage between 19 and 20;
STUDENTID   STUDENTNAME           STUAGE   CLASSID
---------   -----------           ------   ------
2009001     黎明                    20       101
2009002     赵亮                    19       101
2009003     李晓                    20       102

SQL> select * from tblstudent where stuage in (19,20);
STUDENTID   STUDENTNAME           STUAGE   CLASSID
---------   -----------           ------   ------
2009001     黎明                    20       101
2009002     赵亮                    19       101
2009003     李晓                    20       102
```

【例】 查询姓名为空的学生信息。

```
SQL> select * from tblstudent where studentname is null;
STUDENTID   STUDENTNAME           STUAGE   CLASSID
---------   -----------           ------   ------
2009005                             22       101
```

【例】 查询年龄大于等于 19 且姓"李"的学生信息。

```
SQL> select * from tblstudent where stuage >= 19 and studentname like '李%';
STUDENTID   STUDENTNAME           STUAGE   CLASSID
---------   -----------           ------   ------
2009003     李晓                    20       102
```

(7) 分组函数与分组查询

Oracle 数据库中包含一组分组函数，可对一组行进行数据统计，表 9-2 列出了常用的分组函数，这些函数都返回一个 NUMBER 类型的值，且计算过程中忽略空值。

表 9-2 分组函数

函　　数	说　　明
avg(x)	返回 x 列的平均值
min(x)	返回 x 列的最小值
max(x)	返回 x 列的最大值
count(x)	返回包含 x 列的行数
sum(x)	返回 x 列的和
median(x)	返回 x 列的中间值

【例】 分别根据 studentid 和 studentname 统计数据行数。

```
SQL> select * from tblstudent;
STUDENTID  STUDENTNAME              STUAGE    CLASSID
---------  -----------------------  --------  --------
2009001    黎明                       20        101
2009002    赵亮                       19        101
2009003    李晓                       20        102
2009004    李晓芳                               102
2009005                              22        101

SQL> select count(studentid) from tblstudent;
COUNT(STUDENTID)
----------------
               5

SQL> select count(studentname) from tblstudent;
COUNT(STUDENTNAME)
------------------
                 4
```

上面两个查询分别返回 5 和 4，表中实际包含 5 行数据，由于第 5 个学生姓名为空，在根据 studentname 统计时忽略空值，所以返回 4。

【例】 计算班级中学生的最大年龄、最小年龄、中间年龄和平均年龄。

```
SQL> select max(stuage),min(stuage),median(stuage),avg(stuage) from tblstudent;
MAX(STUAGE)  MIN(STUAGE)  MEDIAN(STUAGE)  AVG(STUAGE)
-----------  -----------  --------------  -----------
         22           19              20        20.25
```

分组查询是指根据一个或多个列值将表中数据行进行分组，然后分别计算每组的统计数据。比如统计各个班级学生的最大年龄、最小年龄，这时就需要根据班级来将学生分组，然后分别求出每个班级中学生的最大年龄、最小年龄。

分组查询使用 GROUP BY 子句，GROUP BY 子句后面跟分组列名，可以是一列或多列，查询输出结果中可以包含分组列，除分组列外其他列需要使用分组函数。

【例】 计算每个班级学生的最大年龄和平均年龄。

```
SQL> select classid,max(stuage),avg(stuage) from tblstudent group by classid;
CLASSID  MAX(STUAGE)  AVG(STUAGE)
-------  -----------  -----------
    102           20           20
    101           22    20.3333333
```

【例】 计算每个班级的学生个数。

```
SQL> select classid,count(*) from tblstudent group by classid;
CLASSID      COUNT(*)
-------      ----------
    102             2
    101             3
```

【例】 计算每个班级不同年龄的学生个数。

```
SQL> select classid,stuage,count(*) from tblstudent group by classid,stuage;
CLASSID      STUAGE      COUNT(*)
-------      ----------  ----------
    101          20            1
    102                        1
    101          19            1
    102          20            1
    101          22            1
```

HAVING 子句可以对分组后的数据再进行过滤,放在 GROUP BY 子句后,且必须与 GROUP BY 子句一起使用,但 GROUP BY 子句可以单独使用。

【例】 查询班级学生人数大于 2 的班级及学生数。

```
SQL> select classid,count(*) from tblstudent group by classid having count(*)>2;

CLASSID      COUNT(*)
-------      ----------
    101             3
```

(8) 数据排序

使用 ORDER BY 子句可以对查询出来的数据进行排序,可以指定一列或多列,默认是升序 ASC,也可指定降序 DESC。

【例】 按年龄由小到大显示学生信息。

```
SQL> select * from tblstudent order by stuage;
STUDENTID   STUDENTNAME         STUAGE      CLASSID
---------   --------------      --------    --------
2009002     赵亮                   19           101
2009001     黎明                   20           101
2009003     李晓                   20           102
2009005     李晓                   22           101
2009004     李晓芳                              102
```

【例】 按年龄由大到小显示学生信息。

```
SQL> select * from tblstudent order by stuage desc;
STUDENTID   STUDENTNAME            STUAGE     CLASSID
---------   -------------------    ------     -------
2009004     李晓芳                             102
2009005                            22         101
2009001     黎明                   20         101
2009003     李晓                   20         102
2009002     赵亮                   19         101
```

【例】 按年龄和学号进行排序，年龄为降序，学号升序。

```
SQL> select * from tblstudent order by stuage desc, studentid;
STUDENTID   STUDENTNAME            STUAGE     CLASSID
---------   -------------------    ------     -------
2009004     李晓芳                             102
2009005                            22         101
2009001     黎明                   20         101
2009003     李晓                   20         102
2009002     赵亮                   19         101
```

在 ORDER BY 子句中，多列排序分别指定升序或降序，如未明确声明默认都是升序，排序时也可根据列的次序进行排序。

【例】 根据第 4 列排序。

```
SQL> select * from tblstudent order by 4;
STUDENTID   STUDENTNAME            STUAGE     CLASSID
---------   -------------------    ------     -------
2009001     黎明                   20         101
2009002     赵亮                   19         101
2009005                            22         101
2009004     李晓芳                             102
2009003     李晓                   20         102
```

9.3.2 连接查询

关系数据库中进行表设计时，通常采用数据库三范式来规范表的设计，消除数据冗余，这样原本一个表可以存储的数据经过规范化后通常拆成几个表进行存储，查询时需要从几个表中取数据，这时的查询就是连接查询。下面以员工表为例说明连接查询。

规范化前的表如下。

```
员工表 emp1:
部门编号    部门名称      部门地址      员工编号    员工名称      员工联系电话    员工年龄
deptid      deptname      addr          empid       empname       phonenum        age
--------    --------      --------      --------    --------      --------        --------
```

1001	销售	北京	2012001	李媛	13500234521	30
1001	销售	北京	2012002	张晓丽	13634123401	39
1002	研发	大连	2012003	赵亮	13667676675	40
1002	研发	大连	2012004	秦琦	13823452348	28
1002	研发	大连	2012005	孙不二	13923423452	20

如果查询所有的员工姓名及所在部门名称，直接从表 emp1 中查询即可：

```
select empname, deptname from emp1;
```

规范化后的表如下。

部门表 dept2：

部门编号 deptid	部门名称 deptname	部门地址 addr
1001	销售	北京
1002	研发	大连

员工表 emp2：

员工编号 empid	员工名称 empname	员工联系电话 phonenum	员工年龄 age	部门编号 deptid
2012001	李媛	13500234521	30	1001
2012002	张晓丽	13634123401	39	1001
2012003	赵亮	13667676675	40	1002
2012004	秦琦	13823452348	28	1002
2012005	孙不二	13923423452	20	1002

如果查询所有的员工姓名及所在部门名称，需要表 dept2 和 emp2 连接：

```
select emp2.empname, dept2.deptname
from emp2,dept2
where emp2.deptid = dept2.deptid;
```

1）使用表别名

在多个表存在相同列名时，需要在列名前增加表名来明确指定列。在上面的查询中指定列名时，在列前面都加了表名，查询列表 emp2.empname、dept2.deptname 和 WHERE 条件 emp2.deptid=dept2.deptid 中都在列名前指定了表名，如果查询列较多且表名较长时，反复输入表名将会很烦琐。更好的办法是在 FROM 子句中定义表的别名来表示该表，在查询中的其他地方都通过别名引用该表，比如上面的语句修改后为：

```
select e.empname, d.deptname
from emp2 e, dept2 d
where e.deptid = d.deptid;
```

通过别名引用表，会大大提高 SQL 语句的可读性。

2）笛卡儿积

多表查询时，如果没有连接条件，则一个表的所有行将会和另一个表的所有行都进行连接，连接的结果称为笛卡儿积，比如上面的 dept2 表和 emp2 表连接没有指定连接条件，将会出现下面的结果：

```
select e.empname, d.deptname from emp2 e, dept2 d;
EMPNAME    DEPTNAME
---------  ---------
李媛        销售
张晓丽      销售
赵亮        销售
秦琦        销售
孙不二      销售
李媛        研发
张晓丽      研发
赵亮        研发
秦琦        研发
孙不二      研发
```

一共出现 10 行数据，dept2 表包含两个部门，emp2 包含 5 个员工，结果是全部连接 2×5＝10 行。

3) 内连接

内连接（inner join）是指只有当连接中的列满足连接条件时才会返回一行数据，也就是只返回符合条件的数据，内连接是连接查询中最常用的连接方式。连接条件可以是等值连接（使用等号＝），也可以是不等值连接（除等号外）。

下面以员工工资表、部门表和工资税率表为例演示内连接。

```
员工表 salary:
empid      empname    salary     deptid
---------  ---------  ---------  ---------
2012001    李媛        3500       1001
2012002    张晓丽      6000       1001
2012003    赵亮        4000       1002
2012004    秦琦        5000       1002
2012005    孙不二      5500       1002

部门表 dept:
deptid     deptname
---------  ---------
1001       销售
1002       研发

工资税率表 tax:
low        high       rate       addon
-----      -----      -----      ---------
0          2000       0          0
2001       3000       0.1        0
3001       5000       0.15       100
5001       20000      0.2        400
```

【例】 查询员工姓名及所在部门名称。

```
SQL> select e.empname,d.deptname
       from salary e,dept d
       where e.deptid = d.deptid;
EMPNAME    DEPTNAME
-------    --------
李媛        销售
张晓丽      销售
赵亮        研发
秦琦        研发
孙不二      研发
```

【例】 查询员工姓名、工资及符合的税率信息。

```
SQL> select e.empname,e.salary,t.low,t.high,t.rate,t.addon
       from salary e, tax t
       where e.salary > t.low and  e.salary <= t.high;

EMPNAME    SALARY    LOW     HIGH     RATE    ADDON
-------    ------    ----    -----    ----    -----
李媛        3500      3000    5000     .15     100
赵亮        4000      3000    5000     .15     100
秦琦        5000      3000    5000     .15     100
孙不二      5500      5000    20000    .2      400
张晓丽      6000      5000    20000    .2      400
```

【例】 查询员工姓名及所应缴纳的个人所得税。

```
SQL> select e.empname,(e.salary - t.low) * t.rate + t.addon tax
       from salary e, tax t
       where e.salary > t.low and  e.salary <= t.high;
EMPNAME    TAX
-------    ---
李媛        175
赵亮        250
秦琦        400
孙不二      500
张晓丽      600
```

【例】 查询员工姓名、所在部门、缴纳的个人所得税。

```
SQL> select e.empname,d.deptname, (e.salary - t.low) * t.rate + t.addon tax
       from salary e, tax t, dept d
       where e.deptid = d.deptid and e.salary > t.low and  e.salary <= t.high;
```

```
EMPNAME     DEPTNAME       TAX
--------    ------------   ----------
李嫒         销售           175
张晓丽       销售           600
赵亮         研发           250
秦琦         研发           400
孙不二       研发           500
```

【例】 查询部门名称、部门员工缴纳个人所得税总额。

```
SQL > select d.deptname, sum((e.salary - t.low) * t.rate + t.addon) totaltax
        from salary e, tax t, dept d
        where e.deptid = d.deptid and e.salary > t.low and  e.salary <= t.high
        group by d.deptname;
DEPTNAME    TOTALTAX
--------    ----------
销售        775
研发        1150
```

Oracle 内连接语法基于 ANSI SQL/86 标准，在开发 Oracle 9i 时，还实现了 SQL/92 标准的连接语法，SQL/92 语法引入了 INNER JOIN 和 ON 子句分别指定要连接的表和连接条件，语法如下：

```
SELECT column_list
FROM table_name
INNER JOIN table_name ON conditioin
[ INNER JOIN table_name ON conditioin ]
```

上面的 SQL 语句：

```
select e.empname, d.deptname
from salary e, dept d
where e.deptid = d.deptid;
```

使用 SQL/92 语法重写后为：

```
select e.empname, d.deptname
from salary e
inner join dept d on e.deptid = d.deptid;
```

SQL/92 标准可以使用 USING 子句对连接条件进一步进行简化，但要求必须是等值连接且连接的列名必须相同，上面的 SQL 可修改为：

```
select e.empname, d.deptname
from salary e
inner join dept d using(deptid);
```

4）外连接

外连接（outer join）是指表连接时除了返回符合连接条件的数据外，同时返回不符合连接条件的数据。外连接分为左外连接(left outer join)、右外连接(right outer join)和全外连接(full outer join)。

左外连接：是以左侧表为主，除返回符合连接条件数据外，返回左侧表不符合连接条件的数据。

右外连接：是以右侧表为主，除返回符合连接条件数据外，返回右侧表不符合连接条件的数据。

全外连接：除返回符合连接条件数据外，同时返回所有不符合连接条件的数据。

Oracle中特有的外连接操作符是(＋)，可以使用该连接符执行一个外连接，可以实现左外连接和右外连接，但不能实现全外连接，除了该操作符外可以使用SQL/92语法实现所有的外连接。

下面以教师表和院系表为例演示外连接。

```
教师表 teacher:
tid        tname        did
-----      -----        ----
1001       李燕         1
1002       赵刚         2
1003       李晓利

院系表 department:
did        dname
-----      ------------------
1          计算机分院
2          管理分院
3          外语分院
```

【例】 查询院系名称及院系教师名称，包括没有教师的院系名称(左外连接)。

```
Oracle 语法：
SQL> select d.dname, t.tname
       from department d, teacher t
       where d.did = t.did( + );
DNAME           TNAME
--------        ----------
计算机分院       李燕
管理分院         赵刚
外语分院


SQL/92 标准语法：
SQL> select d.dname, t.tname
       from department d
       left outer join teacher t on d.did = t.did;
DNAME           TNAME
--------        ----------
计算机分院       李燕
管理分院         赵刚
外语分院
```

【例】 查询院系名称及院系教师名称,包括所有没分配院系的教师名称(右外连接)。

```
Oracle 语法:
SQL> select d.dname,t.tname
        from department d, teacher t
        where d.did(+) = t.did;
DNAME        TNAME
--------     ----------
计算机分院    李燕
管理分院      赵刚
              李晓利

SQL/92 标准语法:
SQL> select d.dname,t.tname
        from department d
        right outer join teacher t on d.did = t.did;
DNAME        TNAME
--------     ----------
计算机分院    李燕
管理分院      赵刚
              李晓利
```

【例】 查询院系名称及院系教师名称,包括没有教师的院系名称以及所有没分配院系的教师名称(全外连接)。

```
SQL/92 标准语法:
SQL> select d.dname,t.tname
        from department d
        full outer join teacher t on d.did = t.did;
DNAME        TNAME
--------     ----------
计算机分院    李燕
管理分院      赵刚
              李晓利
外语分院
```

5) 自连接

自连接是指对同一个表的连接,算是内连接的一个特例。要执行一个内连接,必须使用不同的别名来区分,下面以员工表为例来演示内连接。

```
员工表 emp:
eid       ename      mid
-------   -------    --------
1001      李燕
1002      赵刚       1001
1003      李晓利     1002
```

员工表包含员工编号 eid、员工名称 ename 以及上级管理者的编号 mid。

【例】 查询每名员工的姓名及上级管理者的姓名。

```
SQL> select e1.ename 员工, e2.ename 管理者
     from emp e1, emp e2
     where e1.mid = e2.eid( + );
员工         管理者
-------      -------
赵刚         李燕
李晓利       赵刚
```

9.3.3 子查询

前面使用的查询语句都只包含一条 SELECT 语句,而有些情况下依靠单条 SELECT 语句无法完成查询要求,这时需要在 SELECT 语句内部嵌入另外的 SELECT 语句,这条嵌入的 SELECT 语句称为子查询(SUBQUERY),子查询除了可以应用到 SELECT 语句中,也可以应用到 INSERT、UPDATE 和 DELETE 语句中。

子查询分类有如下几种类型。

1) 单行子查询

只返回一行一列数据的子查询称为单行子查询。外部查询可使用比较运算符=、>、>=、<、<=、<>。

2) 多行子查询

返回多行单列数据的子查询称为多行子查询。外部查询可使用 in、not in、any 或 all 操作符。

3) 多列子查询

返回单行或多行多列数据的子查询称为多列子查询。其中单行多列子查询操作类似单行子查询,多行多列子查询操作类似多行子查询。

4) 关联子查询

子查询引用外部查询中包含的一列或多列,查询的执行依赖外部查询。针对每行外部查询数据,都将执行一次子查询。外部查询可使用 exists、not exists 操作符。

下面以学生表、课程表和成绩表为例演示子查询。

```
学生表 student:
sid          sname        sage        class
-------      -------      ------      --------
S001         李颖         19          1班
S002         张良         20          1班
S003         孙小卉       20          2班

课程表 course:
cid          cname
---          --------
C01          数据库
C02          操作系统
```

成绩表 score:

```
sid         cid        score
-------     -----      -------------
S001        C01        65
S002        C01        80
S003        C01        85
S001        C02        60
S002        C02        90
S003        C02        70
```

1. 单行子查询

单行子查询不向外部的 SQL 语句返回结果,或者只返回一行。单行子查询可应用于 SELECT 语句的 WHERE 子句、FROM 子句或 HAVING 子句中。

1) 在 WHERE 子句中使用

子查询作为条件判断的一方,位于小括号中(…)。

【例】 查询年龄最小的学生信息。

```
SQL> select * from student
     where sage = (select min(sage) from student);
SID        SNAME     SAGE    CLASS
-------    ------    ----    -------------
S001       李颖      19      1班
```

上面查询中通过子查询"select min(sage) from student"获取学生中的最小年龄,然后再用于查询条件比较。

【例】 查询大于最小年龄的学生信息。

```
SQL> select * from student
     where sage >(select min(sage) from student);
SID        SNAME     SAGE    CLASS
-------    ------    ----    -------------
S002       张良      20      1班
S003       孙小卉    20      2班
```

2) 在 FROM 子句中使用

用于 FROM 子句中的子查询也称为内联视图(inline view)。

【例】 查询年龄不小于 20 的所有学生姓名。

```
SQL> select sname from (select * from student where sage >= 20);
SNAME
-------------
张良
孙小卉
```

3) 在 HAVING 子句中使用

【例】 查询平均成绩大于总平均成绩的学生学号和平均成绩。

```
SQL> select sid,avg(score)
    from score
    group by sid
    having avg(score)>(select avg(score) from score);
SID        AVG(SCORE)
----       ----------
S002       85
S003       77.5
```

2. 多行子查询

多行子查询通常返回一条或多条记录,多行子查询结果用于 WHERE 语句时,判断可以使用 IN、ANY 或 ALL 操作符。

【例】 查询课程成绩高于 80 的学生信息。

```
SQL> select * from student
    where sid in (select sid from score where score>80);
SID        SNAME      SAGE    CLASS
-------    -----      ---     -----
S003       孙小卉     20      2班
S002       张良       20      1班
```

通过子查询"select sid from score where score>80"获取所有成绩高于 80 的学号,然后运用集合操作符 IN 从学生表中获取符合条件的学生信息。

【例】 查询年龄大于等于所有班级平均年龄的学生信息。

```
SQL> select * from student
    where sage >= all (select avg(sage) from student group by class);
SID        SNAME      SAGE    CLASS
-------    -----      ---     -----
S002       张良       20      1班
S003       孙小卉     20      2班
```

上面表中共有三名学生,分属两个班级,两个班级的平均年龄分别是 19.5 和 20,上面查询结果中的两个学生年龄为 20,大于等于两个班级的平均年龄。操作符 >=ALL 等同于大于等于子查询结果中最大值,如果是<=ALL 等同于小于等于子查询结果中最小值。

【例】 查询单科成绩小于任意课程平均成绩的学号、课程号和成绩。

```
SQL> select sid,cid,score from score
    where score < any (select avg(score) from score group by cid);
SID        CID        SCORE
----       ----       -----
S001       C01        65
S001       C02        60
S003       C02        70
```

上面子查询"select avg(score) from score group by cid"返回每门课程的平均成绩,只要成绩比其中的一个小即满足条件。

3. 多列子查询

前面的子查询结果无论是单行还是多行返回的都是单列值,多列子查询指子查询返回结果包含多列,如果是单行可用操作符=、<>,如果是多行可用操作符 IN,数值比较时是多列比较,多个列用小括号括起,列之间用逗号隔开。

【例】 查询每门课分数最低的学号、课程号和成绩。

```
SQL> select sid,cid,score from score
    where (cid,score) in
    (select cid,min(score) from score  group by cid);
SID        CID     SCORE
-------    -----   ----------
S001       C01     65
S001       C02     60
```

【例】 查询每门课分数最低的学生姓名、课程名称和成绩。

```
SQL> select s.sname,c.cname,sc.score from student s,course c,score sc
    where s.sid = sc.sid
    and c.cid = sc.cid
    and (sc.cid,sc.score) in
    (select cid,min(score) from score group by cid);
SNAME    CNAME          SCORE
-------  ------------   ----
李颖     操作系统        60
李颖     数据库          65
```

4. 关联子查询

查询中不需要外部父查询信息,称为非关联子查询。如子查询中引用外部父查询信息,称为关联子查询。非关联子查询只在运行外部查询时运行一次,而关联子查询对于外部查询中的每一行都会运行一次。

【例】 查询成绩大于本门课程平均成绩的学号、课程号和成绩。

```
SQL> select * from score outer
    where score >
    (select avg(score) from score inner where inner.cid = outer.cid);
SID        CID          SCORE
-------    ----------   ----
S002       C01          80
S003       C01          85
S002       C02          90
```

外部查询中的每一行成绩,都将执行一次子查询,以课程号作为子查询运行的条件求出该门课的平均成绩,然后判断当前行成绩是否大于子查询计算出的平均成绩。

如果只是用于判断子查询返回行的存在性,可使用 EXISTS 操作符,虽然该操作符可用于非关联子查询,但更多情况下用于关联子查询。EXISTS 操作符判断是否存在,NOT EXISTS 操作符判断是否不存在。

下面以员工表 emp 为例演示操作符 EXISTS 的用法。

```
员工表 emp:
eid        ename      mid
-------    -------    ---------
1001       李燕
1002       赵刚        1001
1003       李晓利      1002
```

【例】 查询负责管理其他员工的员工姓名。

```
SQL> select ename from emp e1
     where exists (select 1 from emp e2 where e2.mid = e1.eid);
ENAME
--------------
李燕
赵刚
```

EXISTS 操作符只检查返回行的存在性,因此子查询不必返回一行,返回一个常量即可,可提高查询性能。如果子查询返回一行或多行,EXISTS 返回 TRUE,否则返回 FALSE。

【例】 查询不负责管理其他员工的员工姓名。

```
SQL> select ename from emp e1
     where not exists (select 1 from emp e2 where e2.mid = e1.eid);
ENAME
--------------
李晓利
```

用 IN 操作符实现上面的查询,如下所示:

```
SQL> select ename from emp e1
     where eid in (select mid from emp );
ENAME
--------------
李燕
赵刚


SQL> select ename from emp e1
     where eid not in (select mid from emp );
no rows selected
```

上面的子查询"select mid from emp"结果包括 1001、1002 和空值，使用 IN 或 NOT IN 操作符判断某个值是否在子查询结果中时，任何值与空值 NULL 比较都返回 FALSE。员工号 1001 和 1002 在子查询结果中，使用 IN 判断时返回 TURE，1003 不在子查询结果中，与空值比较返回 FALSE，所以第一个查询结果正确，返回两行。第二个查询使用 NOT IN 判断，员工号 1001 和 1002 在子查询结果中，使用 NOT IN 判断时返回 FALSE，1003 不在子查询结果中，但子查询结果包含空值，使用 NOT IN 判断返回 FALSE，这时没有任何结果返回，查询结果不正确。

5. 编写包含子查询的 DML 语句和 DDL 语句

1) 在 DDL 语句中使用子查询

【例】 创建新表，使用表 STUDENT 结构和数据。

```
SQL> create table temp1
    as select * from student;
```

【例】 创建新表，使用表 STUDENT 结构。

```
SQL> create table temp2
    as select * from student where 1 = 2;
```

2) 在 DML 语句中使用子查询

【例】 将年龄最小的学生年龄替换成平均年龄。

```
SQL> update student set sage =
    (select avg(sage) from student)
    where sage = (select min(sage) from student);
```

【例】 将 STUDENT 表中数据插入到表 TEMP2 中。

```
SQL> insert into temp2
    select * from student;
```

【例】 删除年龄最小的学生信息。

```
SQL> delete from student
    where sage = ( select min(sage) from student ) ;
```

9.3.4 高级查询

1. 集合操作

集合操作可以对多个查询结果集进行运算，但前提是多个查询结果集包含相同数量的列以及相同的列类型。表 9-3 列出了集合操作符。

表 9-3　集合操作符

操　作　符	说　　　明
UNION ALL	返回各个查询检索出的所有行,包括重复行
UNION	返回各个查询检索出的所有行,不包括重复行
INTERSECT	返回两个查询检索的共有行
MINUS	返回第一个查询减去第二个查询返回的行

下面以表 PROD1 和 PROD2 为例演示集合操作。

```
产品表 PROD1:
产品编号    产品名称    品牌
---------   ---------   -------------
1           电视        三星
2           冰箱        海尔
3           洗衣机      西门子

产品表 PROD2:
产品编号    产品名称    品牌
---------   ---------   -------------
1           洗衣机      西门子
2           热水器      西门子
```

1) UNION ALL 操作

UNION ALL 操作只合并结果集,不去除重复数据。

【例】　查询所有产品。

```
SQL> select pname,producer from prod1
    union all
    select pname,producer from prod2;
PNAME       PRODUCER
---------   --------------------
电视        三星
冰箱        海尔
洗衣机      西门子
洗衣机      西门子
热水器      西门子
```

2) UNION 操作

UNION 操作合并结果集后去除重复数据。

【例】　查询所有产品。

```
SQL> select pname,producer from prod1
    union
    select pname,producer from prod2;
```

```
PNAME      PRODUCER
--------   --------------------
电视        三星
冰箱        海尔
洗衣机      西门子
热水器      西门子
```

3) INTERSECT 操作

INTERSECT 操作返回结果集间的交集,也就是共有部分。

【例】 查询两个产品表的共有产品。

```
SQL> select pname,producer from prod1
    intersect
    select pname,producer from prod2;

PNAME      PRODUCER
--------   --------------------
洗衣机      西门子
```

4) MINUS 操作

MINUS 操作计算结果集间的差,返回第一个结果集去掉共有部分的数据。

【例】 查询两个产品表 PROD1 与 PROD2 的差。

```
SQL> select pname,producer from prod1
    minus
    select pname,producer from prod2;

PNAME      PRODUCER
--------   --------------------
电视        三星
冰箱        海尔
```

2. CASE 语句

CASE 语句可以应用到 SQL 语句中,实现 if-then-else 的逻辑,与函数 DECODE 类似,Oracle 数据库从 9i 开始支持,CASE 语句已经成为 SQL/92 标准的一部分。CASE 语句有以下两种形式。

1) 简单 CASE 语句

使用表达式确定返回值,语法如下:

```
CASE search_expression
    WHEN expression1 THEN result1
    WHEN expression2 THEN result2
    ...
    ELSE default_result
END
```

【例】 查询学生表,性别是 1 输出"男",2 输出"女",其他输出"未知"。

```
SQL> select sname,
            case gender
                when 1 then '男'
                when 2 then '女'
                else '未知'
            end
     from student;
SNAME      GENDER
--------   --------------------
李颖        男
张良        女
孙小卉      未知
```

2) CASE 搜索语句

使用条件表达式确定返回值,语法如下:

```
CASE true|false
    WHEN condition1 THEN result1
    WHEN condition2 THEN result2
    ELSE default_result
END
```

如果 CASE true(CASE 语句默认为 CASE true),下列条件为真时执行相应语句,如果 CASE false,下列条件为假时执行相应语句。

【例】 查询学生表,性别是 1 输出"男",2 输出"女",其他输出"未知"。

```
SQL> select sname,
            case
                when gender = 1 then '男'
                when gender = 2 then '女'
                else '未知'
            end
     from student;
SNAME      GENDER
--------   --------------------
李颖        男
张良        女
孙小卉      未知
```

3. 层次查询

人们可以经常见到有层次的数据,比如企业员工信息,每名员工都有上级领导,而上级领导同样也是员工,普通员工与领导信息共存于一张表,这时就需要层次查询来体现出层次关系。下面以员工表为例讲解层次查询。

员工表 emp3:
eid   ename   mid

```
-------    -------    ---------
1001       李燕
1002       赵刚       1001
1003       李晓利     1001
1004       李飞       1002
1005       张力       1003
```

员工的层次关系如图 9-1 所示。

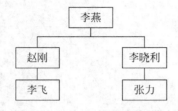

图 9-1　员工关系图

层次查询语法如下：

```
SELECT [LEVEL], select_list
FROM table_name
START WITH start_condition
CONNECT BY PRIOR prior_condition
```

属性说明：

1) LEVEL

是一个伪列，代表层次查询的层数，根节点层数为 1。

2) start_condition

定义层次查询的起点，比如从员工号 1002 开始，就写成 eid='1002'。

3) prior_condition

定义层次查询父行和子行的关系，比如写成 eid=mid，也就是从起点开始从上向下遍历，反过来写成 mid=eid 则做从下向上遍历。

【例】从员工 1001 开始做从上向下遍历。

```
SQL> select level, eid, ename from emp
    start with eid = '1001'
    connect by prior eid = mid
    order by level;

LEVEL  EID   ENAME
-----  ---   --------------------
1      1001  李燕
2      1003  李晓利
2      1002  赵刚
3      1005  张力
3      1004  李飞
```

【例】 从员工1002开始做从上向下遍历。

```
SQL> select level,eid,ename from emp
    start with eid = '1002'
    connect by prior eid = mid;

LEVEL  EID   ENAME
-----  ---   ------------------
   1   1002  赵刚
   2   1004  李飞
```

【例】 从员工1005开始做从下向上遍历。

```
SQL> select level,eid,ename from emp
    start with eid = '1005'
    connect by prior mid = eid;

LEVEL  EID   ENAME
-----  ---   ------------------
   1   1005  张力
   2   1003  李晓利
   3   1001  李燕
```

【例】 从员工1001开始做从上向下遍历,利用LEVEL格式化结果。

```
SQL> select level,eid,lpad('*',(level-1)*2,'*')||ename ename from emp
    start with eid = '1001'
    connect by prior eid = mid
    order by level;

LEVEL  EID   ENAME
-----  ---   ------------------
   1   1001  李燕
   2   1003  **李晓利
   2   1002  **赵刚
   3   1005  ****张力
   3   1004  ****李飞
```

4. 扩展分组语句

GROUP BY 子句中可以使用 ROLLUP 或 CUBE,ROLLUP 可以为每个分组返回一个小计以及为所有分组返回一个总计记录,CUBE 可以返回每个分组列组合的小计记录以及在末尾加上总计记录。下面以学生表为例讲解 ROLLUP 和 CUBE 的使用。

```
学生表 student:
name      age     gender
------    ---     ----------------
李燕      20      女
赵刚      20      女
李晓利    20      男
```

| 李飞 | 21 | 男 |
| 张力 | 21 | 男 |

1）使用 ROLLUP 进行分组统计

ROLLUP 分组统计按照分组列从右向左依次减一列进行统计，比如分组列为(c1,c2,c3)，则分组统计将分别按照(c1,c2,c3)、(c1,c2)、(c1)进行统计，最后增加总计。

【例】 按年龄统计学生人数。

```
SQL> select age, count(*) from student group by rollup(age);

AGE   COUNT(*)
---   --------
20        3
21        2
          5
```

【例】 按年龄和性别统计学生人数。

```
SQL> select age,gender,count(*) from student
     group by rollup(age,gender);

AGE   GENDER   COUNT(*)
---   ------   --------
20    女           2
20    男           1
20                3
21    男           2
21                2
                  5
```

2）使用 CUBE 进行分组统计

CUBE 分组统计对分组列进行组合分别统计，比如分组列为(c1,c2,c3)，则分组统计将分别按照(c1,c2,c3)、(c1,c2)、(c1,c3)、(c2,c3)、(c1)、(c2)、(c3)进行统计，最后增加总计。

【例】 按年龄和性别统计学生人数。

```
SQL> select age, gender,count(*) from student
     group by cube(age,gender)
     order by age,gender;

AGE   GENDER   COUNT(*)
---   ------   --------
20    女           2
20    男           1
20                3
21    男           2
21                2
      女           2
      男           3
                  5
```

3) 使用 GROUPING 函数

GROUPING 函数可以接受一分组列,返回 0 或 1,如果该列值为空,返回 1,否则返回 0。GROUPING 函数只能在 ROLLUP 和 CUBE 中使用,在格式化输出时非常有用。

【例】 按年龄和性别统计学生人数,显示 GROUPING 函数值。

```
SQL> select grouping(age),grouping(gender),age,gender,count(*) from student
    group by rollup(age,gender);
GROUPING(AGE)  GROUPING(GENDER)  AGE  GENDER  COUNT(*)
-------------  ----------------  ---  ------  --------
            0                 0   20  女             2
            0                 0   20  男             1
            0                 1   20                 3
            0                 0   21  男             2
            0                 1   21                 2
            1                 1                      5
```

【例】 按年龄和性别统计学生人数,根据 GROUPING 函数值转换输出信息。

```
SQL> select
        case grouping(age)
            when 1 then '所有年龄'
            else ''||age
        end as age,
        case grouping(gender)
            when 1 then '所有性别'
            else gender
        end as gender,
        count(*) total
    from student
    group by rollup(age,gender);

AGE       GENDER     TOTAL
------    --------   --------
20        女          2
20        男          1
20        所有性别    3
21        男          2
21        所有性别    2
所有年龄  所有性别    5
```

9.4 事务控制语句

9.4.1 Oracle 事务开始结束条件

数据库事务就是一组 SQL 语句,这组 SQL 语句是一个逻辑工作单元。可以认为事务是一组不可分割的 SQL 语句,其结果应该作为一个整体永久性地修改数据库的内容,或者作为一个整体取消对数据库的修改。

Oracle 数据库事务的开始条件：上一个事务结束后执行任何一条 DML 语句即开始一个新事务。

Oracle 数据库事务在下列事件之一发生时，事务结束。

(1) 执行 commit 或 rollback 语句。

(2) 执行一条 DDL 语句，这种情况下自动执行 commit。

(3) 执行一条 DCL 语句，这种情况下自动执行 commit。

(4) 断开与数据库连接。SQL*Plus 中输入 exit 命令，自动执行 commit；如意外终止 SQL*Plus 就会自动执行 rollback。

9.4.2 事务控制语句

Oracle 数据库事务控制语句包括提交 COMMIT、回滚 ROLLBACK 和保存点 SAVEPOINT。

1. COMMIT 语句

COMMIT 语句用来保证事务修改内容永久保存，当用户执行 COMMIT 语句时，Oracle 数据库将首先保存相应的 redo 数据到日志文件（LGWR 进程将 LOG BUFFER 中数据写出），保证修改的数据永久有效。

2. ROLLBACK 语句

ROLLBACK 语句用来撤销事务修改内容，当用户执行 ROLLBACK 语句时，Oracle 数据库利用保存在 UNDO 段中的前镜像数据来还原变化，恢复成事务修改前的状态。

3. SAVEPOINT 语句

SAVEPOINT 语句用来在事务中设置保存点，可以控制 ROLLBACK 回滚的范围，如没有保存点将回滚所有数据变化，如有保存点，可指定回滚到保存点位置。

【例】 事务操作。

```
SQL> create table dept(id number);
SQL> insert into dept values(1);
SQL> savepoint s1;
SQL> insert into dept values(2);
SQL> rollback to savepoint s1;
SQL> create table emp(id number);
SQL> rollback;
SQL> select * from dept;
        ID
----------
         1
```

上面的例子中，当回滚到保存点 s1 时，插入的第二条数据被回滚，接下来执行了一条 DDL 语句，自动提交事务，事务结束，之后再 rollback 不能回滚事务，最后保留插入的第一条数据。

小 结

本章首先介绍了 SQL 语句分类,接下来重点讲解了数据操作语句、查询语句和事务控制语句。数据操作语句部分讲解了 INSERT、UPDATE、DELTE 和 MERGE 语句的用法;查询语句部分分别介绍了单表查询、连接查询、子查询和其他高级查询;事务控制语句部分介绍了 Oracle 数据库事务开始结束条件和事务语句用法。

思 考 题

1. SQL 语句有哪些类别?分别举例说明。
2. 创建员工表 emp,插入如下数据,写出插入数据语句。

```
create table emp(
    empid number,
    empname varchar2(20),
    deptname varchar2(20),
    gender varchar2(20),
    age number,
    salary number);
empid     empname    deptname    gender    age    salary
-------   --------   ---------   -------   ---   -------
101       张三        销售         男        25    3000
102       王五        销售         男              4000
201       赵六        开发         女        29    5000
202       张小二      开发         女              6000
```

3. 根据思考题 2 中的表 emp,完成如下查询。
(1) 查询所有员工名称、部门名称和工资。
(2) 查询所有男员工的姓名和工资。
(3) 查询所有年龄为空且姓张的员工信息。
(4) 查询所有员工的平均工资、最大工资和最小工资。
(5) 查询各部门员工的平均工资、最大工资和最小工资。
(6) 查询部门平均工资大于 4500 元的部门名称和平均工资。
4. 根据思考题 2 中的表 emp,完成如下修改。
(1) 将年龄为空的员工年龄修改成 30。
(2) 将性别为男且年龄大于 25 的员工工资涨 1000 元。
(3) 将工资最低的员工工资修改成所有员工的平均工资。
(4) 删除工资最高的员工信息。
5. 假设存在课程表 course、学生表 student 和成绩表 score,数据如下,完成查询。

```
课程表 course:  couno    couname
                -----    ----------
                c01      操作系统
```

```
                c02      数据库
学生表 student: stuno    stuname     claname       gender    age
                -----    -------     -------       ------    ---
                s1001    赵亮        10软件1班      男        20
                s1002    李涵        10软件1班      男        18
                s1003    王思梅      10软件2班      女        19

成绩表 score:   stuno    couno       score
                -----    -----       -----
                s1001    c01         80
                s1001    c02         75
                s1002    c01         90
                s1002    c02         95
                s1003    c01         90
                s1003    c02         85
```

（1）查询学生名称、所在班级名称、课程名称及课程分数。

（2）查询班级名称、课程名称及班级所有同学该门课程平均分数。

（3）查询班级名称、学生名称、学生所学的所有课程的平均成绩。

（4）查询学生所学所有课程的平均成绩在 80 以上的班级名称、学生名称以及学生所学所有课程的平均成绩。

（5）查询课程名称，以及该门课程的平均分、最高分、最低分。

6．参见思考题 5 中的表，完成下面的查询并写出语句。

（1）查询年龄最小的学生信息。

（2）查询不同性别的年龄最小的学生信息。

（3）查询年龄小于所有学生平均年龄的学生信息。

（4）查询年龄小于同性别学生平均年龄的学生信息。

（5）查询每门课程中得分最低的学生及课程信息，输出信息包括学生姓名、课程姓名和分数。

（6）查询每门课程中得分最低和最高的学生及课程信息，输出信息包括学生姓名、课程名称和分数。

7．Oracle 数据库中事务开始条件和结束条件分别是什么？

8．事务相关语句有哪些？简述其作用。

第 10 章　内 置 函 数

Oracle 数据库提供大量内置函数,熟练掌握函数用法将大大提高 SQL 语句的编写效率和质量。本章将介绍常用的内置函数使用。

10.1　内置函数简介

Oracle 提供了很多内置函数,主要分为以下两类。

1. 分组函数

分组函数对多行数据操作,返回一行输出结果,具体函数介绍参见 9.3.1 节。

2. 单行函数

单行函数只对一行数据操作,对输入的每一行返回一个输出结果。

【例】　转换大写。

```
SQL> select upper('how are you?') from dual;
UPPER('HOWAREYOU?')
----------------------
HOW ARE YOU?
```

DUAL 表称为虚表,用来执行不对任何特定表进行操作的查询,可以做基本运算处理。它只有一列:DUMMY,其数据类型为 VARCHAR2(1)。DUAL 中只有一行数据:'X'。

本章主要介绍单行函数,包括字符型函数、数值型函数、日期型函数、转换函数以及正则表达式函数。

10.2　字 符 函 数

字符函数接受字符类型的参数,参数可以是字面值,也可以是表中的列值,并返回一个结果。表 10-1 列出了常用的字符函数。

表 10-1　字符函数

函　　数	说　　明
ASCII(x)	返回字符 x 的 ASCII 码
CHR(x)	返回 ASCII 码为 x 的字符
CONCAT(x,y)	数据连接,将 y 添加到 x 上返回

续表

函　　数	说　　明
INITCAP(x)	将 x 中的每个单词的首字母转大写,其余字母转小写
INSTR(x,find_string [,start][,occurrence])	在 x 中查找串 find_string,返回位置
LENGTH(x)	计算 x 的字符个数
LOWER(x)	将字母转换成小写
UPPER(x)	将字母转换成大写
LPAD(x,width[,pad_string])	用于在 x 左边补齐,使长度达到 width 指定的长度
RPAD(x,width[,pad_string])	用于在 x 右边补齐,使长度达到 width 指定的长度
LTRIM(x[,trim_string])	用于从 x 左边截去一些字符,默认截去空格
RTRIM(x[,trim_string])	用于从 x 右边截去一些字符,默认截去空格
TRIM([trim_string FROM]x)	用于从 x 两侧截去一些字符,默认截去空格
NVL(x,value)	如果 x 为空,返回 value,否则返回 x
NVL2(x,value1,value2)	如果 x 不空,返回 value1,否则返回 value2
REPLACE(x,search_string,replace_string)	在 x 中查找 search_string,用 replace_string 替换
SUBSTR(x,start[,length])	在 x 中 start 位置开始取 length 指定长度的字符串,如不指定 length,则取 start 位置后的所有字符

1. ASCII(x)

返回字符 x 的 ASCII 码,或字符串 x 首个字母的 ASCII 码。

```
SQL> select ascii('a'), ascii('abc') from dual;
ASCII('A')          ASCII('ABC')
--------------      ----------------
97                  97
```

2. CHR(x)

返回 ASCII 码为 x 的字符,与 ASCII 函数效果是互逆的。

```
SQL> select chr(97) from dual;
CHR(97)
-------
c
```

3. CONCAT(x,y)

数据连接,将 y 添加到 x 上返回。

```
SQL> select concat('Lucy ','where are you going?') from dual;
CONCAT('LUCY','WHEREAREYOUGOING?')
---------------------------------
Lucy where are you going?
```

CONCAT 函数功能与连接符 || 实现的功能相同。

4. INITCAP(x)

将 x 中的每个单词的首字母换成大写,其余字母转成小写。

```
SQL> select initcap('TODAY IS A GOOD DAY. ') from dual;
INITCAP('TODAYISAGOODDAY.')
-----------------------------------------
Today Is A Good Day.
```

5. INSTR(x, find_string [,start][,occurrence])

在 x 中查找串 find_string,返回位置,如没查到返回 0。

参数 start 指定从哪个位置开始查找,该参数默认值为 1。

参数 occurrence 指定第几次出现,该参数默认值为 1。

```
SQL> select instr('Do you like dog? ','do',5,1) from dual;
INSTR('DOYOULIKEDOG?','DO',5,1)
-------------------------------
                             13
```

6. LENGTH(x)

计算 x 的字符个数,如果是数字小数点也计算在内,如果是日期则根据日期格式(nls_date_format)来计算。

```
SQL> select length('abc'),length('工作'),length(100.123) from dual;
LENGTH('ABC') LENGTH('工作') LENGTH(100.123)
------------- -------------- ---------------
            3              2               7
```

7. LOWER(x)

将字母转换成小写。

```
SQL> select lower('ABC') from dual;
LOWER('ABC')
------------
abc
```

8. UPPER(x)

将字母转换成大写。

```
SQL> select upper('Hello') from dual;
UPPER('HELLO')
--------------
HELLO
```

9. LPAD(x, width[, pad_string])

用于在 x 左边补齐空格,使长度达到 width 指定的长度。如果指定 pad_string 则使用该字符串来填充,该参数可指定单个字符也可指定一个字符串。

width 如果小于 x 本身长度,则保留 width 指定长度的数据。
pad_string 如果为字符需用单引号,如果是数字可直接使用。

```
SQL> select lpad('x',5,'#'), lpad('x',5,0) from dual;
LPAD('X',5,'#')  LPAD('X',5,0)
---------------  -------------
####x            0000x
```

10. RPAD(x,width[,pad_string])

在 x 右边补齐,可参照 LPAD 使用。

11. LTRIM(x[,trim_string])

用于从 x 左边截去一些字符,默认截去空格。trim_string 用于指定要截去的字符,可指定单个字符也可指定一个字符串。

```
SQL> select ltrim('a') trim1,ltrim('##a','#') trim2,
ltrim('##abc','##a') trim3 from dual;
TRIM1  TRIM2  TRIM3
-----  -----  -----
a      a      bc
```

12. RTRIM(x[,trim_string])

用于从 x 右边截去一些字符,可参照 LTRIM 使用。

13. TRIM([trim_string FROM]x)

从 x 左右截去一些字符,默认截去空格。其中 trim_string 只能是单个字符。

```
SQL> select trim('abc ') trim1,trim( 'x' from 'xxabcxx') trim2 from dual;
TRIM1    TRIM2
-------  ------
abc      abc
```

14. NVL(x,value)

如果 x 为空,返回 value,否则返回 x。

```
SQL> select * from student;
SID      SNAME     SAGE      CLASS
-------  --------  --------  --------
S001     李颖      19        1班
S002               20        1班
S003     孙小卉              2班

SQL> select sid, nvl(sname,'无名'), nvl(sage,18),class from student;
SID    NVL(SNAME,'无名')    NVL(SAGE,18) CLASS
-----  -----------------    ------------ --------
S001   李颖                 19           1班
S002   无名                 20           1班
S003   孙小卉               18           2班
```

15. NVL2(x, value1, value2)

如 x 不空返回 value1，否则返回 value2。

```
SQL> select nvl2(sname,sname,'无名') from student;
NVL2(SNAME,SNAME,'无名')
------------------------
李颖
无名
孙小卉
```

16. REPLACE(x, search_string, relace_string)

在 x 中查找 search_string，用 replace_string 替换。

```
SQL> select replace('hi,how are you?','hi','hello') from dual;
REPLACE('HI,HOWAREYOU?','HI','HELLO')
-------------------------------------
hello,how are you?
```

17. substr(x, start[, length])

在 x 中 start 位置开始取 length 指定长度的字符串，如不指定 length，则取 start 位置后的所有字符。

```
SQL> select substr('123456789',3,3) from dual;
SUBSTR('123456789',3,3)
-----------------------
345
```

10.3 数字函数

数字函数接受数字类型的参数，参数可以是字面值，也可以是表中的列值，并返回一个结果。表 10-2 列出了常用的数字函数。

表 10-2 数字函数

函　　数	说　　明
ABS(x)	返回 x 的绝对值
FLOOR(x)	返回小于等于 x 的最大整数
CEIL(x)	返回大于等于 x 的最小整数
MOD(x, y)	返回 x 除以 y 的余数
ROUND(x[, y])	返回对 x 进行四舍五入后的值
TRUNC(x[, y])	返回对 x 进行截断后的值
POWER(x[, y])	返回 x 的 y 次幂

1. ABS(x)

用于得到 x 的绝对值。

```
SQL> select abs(-1),abs(100) from dual;
ABS(-1)    ABS(100)
--------   --------
      1         100
```

2. FLOOR(x)

返回小于或等于 x 的最大整数。

```
SQL> select floor(5.1),floor(-5.5) from dual;
FLOOR(5.1)    FLOOR(-5.5)
----------    -----------
         5             -6
```

3. CEIL(x)

返回大于或等于 x 的最小整数。

```
SQL> select ceil(5.1),ceil(-5.5) from dual;
CEIL(5.1)    CEIL(-5.5)
---------    ----------
        6            -5
```

4. MOD(x,y)

返回 x 除以 y 的余数。

```
SQL> select mod(10,3),mod(-8,3) from dual;
MOD(10,3)    MOD(-8,3)
---------    ---------
        1           -2
```

5. ROUND(x[,y])

四舍五入,y 代表小数位数。

```
SQL> select round(56.56),round(56.56,1),round(56.56,-1) from dual;
ROUND(56.56)  ROUND(56.56,1)  ROUND(56.56,-1)
------------  --------------  ---------------
          57            56.6               60
```

6. TRUNC(x[,y])

截断数据,y 代表对第几位小数截断。

```
SQL> select trunc(56.56),trunc(56.56,1),trunc(56.56,-1) from dual;
TRUNC(56.56)  TRUNC(56.56,1)  TRUNC(56.56,-1)
------------  --------------  ---------------
          56            56.5               50
```

7. POWER(x[,y])

返回 x 的 y 次幂。

```
SQL> select power(2,3),power(3,0) from dual;
POWER(2,3)    POWER(3,0)
----------    ----------
         8             1
```

10.4 日期函数

日期函数接受日期类型的参数,参数可以是字面值,也可以是表中的列值,并返回一个结果。表 10-3 列出了常用的日期函数。

表 10-3 日期函数

函 数	说 明
SYSDATE	返回数据库服务器所在操作系统当前时间
ADD_MONTHS(x,y)	返回日期 x 加上 y 个月的结果,y 可以是正数或负数
LAST_DAY(x)	返回日期 x 所在月的最后一天的日期
MONTHS_BETWEEN(x,y)	返回 x 和 y 间隔月数
NEXT_DAY(x,day)	返回日期 x 开始下一个 day 的日期
ROUND(x[,unit])	时间值取整,默认情况下,x 取整为最近的一天
TRUNC(x[,unit])	时间值截断,默认情况下,x 被截为当天的开始时间

1. SYSDATE

返回数据库服务器所在操作系统当前时间。

```
SQL> alter session set nls_date_format = 'yyyy-mm-dd hh24:mi:ss';
Session altered.

SQL> select sysdate from dual;
SYSDATE
-------------------
2013-08-27 06:43:14
```

2. ADD_MONTHS(x,y)

返回日期 x 加上 y 个月的结果,y 可以是正数或负数。

```
SQL> select sysdate,add_months(sysdate,2) from dual;
SYSDATE              ADD_MONTHS(SYSDATE,2)
-------------------  ---------------------
2013-08-27 06:51:23  2013-10-27 06:51:23
```

3. LAST_DAY(x)

返回日期 x 所在月的最后一天的日期。

```
SQL> select sysdate,last_day(sysdate) from dual;
SYSDATE                LAST_DAY(SYSDATE)
-------------------    -------------------
2013-08-27 06:52:08    2013-08-31 06:52:08
```

【例】 查询生日是月末的所有学生信息。

```
SQL> select sname,birth from student where birth = last_day(birth);
SID      SNAME         SAGE   BIRTH
------   -----------   ----   -------------------
S002     张良                 1991-08-31 09:30:00
```

4. MONTHS_BETWEEN(x,y)

返回 x 和 y 间隔月数。

```
SQL> select * from student;
SID      SNAME         SAGE   BIRTH
------   -----------   ----   -------------------
S001     李颖                 1990-10-01 10:01:00
S002     张良                 1991-08-31 09:30:00
S003     孙小卉               1991-02-10 09:30:00

SQL> select sname,months_between(sysdate,birth) from student;
SNAME              MONTHS_BETWEEN(SYSDATE,BIRTH)
-----------------  -----------------------------
李颖                                    274.834503
张良                                    263.867456
孙小卉                                  270.544875
```

【例】 计算学生年龄并更新列 sage。

```
SQL> update student set sage = ceil(months_between(sysdate,birth)/12)
3 rows updated.

SQL> select * from student;
SID      SNAME         SAGE   BIRTH
------   -----------   ----   -------------------
S001     李颖           23    1990-10-01 10:01:00
S002     张良           22    1991-08-31 09:30:00
S003     孙小卉         23    1991-02-10 09:30:00
```

5. NEXT_DAY(x,day)

返回日期 x 开始下一个 day 的日期,day 是一个字符串,比如 Saturday。

```
SQL> select sysdate,to_char(sysdate,'day'),next_day(sysdate,'tuesday') from dual;
SYSDATE        TO_CHAR(SYSDATE,'DAY')      NEXT_DAY(SYSDATE,'TUESDAY')
----------     ----------------------      ---------------------------
2013-08-27     Tuesday                     2013-09-03
```

6. ROUND（x[,unit]）

时间值取整,默认情况下,x 取整为最近的一天。unit 是可选参数,指明要取整单元。

```
SQL> select sysdate from dual;
SYSDATE
-------------------
2013-08-27 06:58:22

SQL> select round(sysdate) from dual;
ROUND(SYSDATE)
-------------------
2013-08-27 00:00:00
SQL> select round(sysdate,'hh24') from dual;
ROUND(SYSDATE,'HH24')
---------------------
2013-08-27 07:00:00

SQL> select round(sysdate,'mm') from dual;
ROUND(SYSDATE,'MM')
-------------------
2013-09-01 00:00:00
```

7. TRUNC（x[,unit]）

时间值截断,默认情况下,x 被截为当天的开始时间,unit 参数同 round 函数中使用方法。

```
SQL> select sysdate from dual;
SYSDATE
-------------------
2013-08-27 06:59:13

SQL> select trunc(sysdate) from dual;
TRUNC(SYSDATE)
-------------------
2013-08-27 00:00:00

SQL> select trunc(sysdate,'hh24') from dual;
TRUNC(SYSDATE,'HH24')
---------------------
2013-08-27 06:00:00

SQL> select trunc(sysdate,'mm') from dual;
TRUNC(SYSDATE,'MM')
-------------------
2013-08-01 00:00:00
```

10.5 转换函数

转换函数可完成数字、字符串和日期间的相互转换，参数可以是字面值，也可以是表中的列值，并返回一个结果。表 10-4 列出了常用的转换函数。

表 10-4 转换函数

函 数	说 明
TO_CHAR(x[,format])	将 x 转换成字符类型
TO_DATE(x[,format])	将 x 转换成日期类型
TO_NUMBER(x[,format])	将 x 装换成数字类型
TRANSLATE(x,from_string,to_string)	在 x 中查找 from_string 中的字符，将其替换成 to_string 中对应的字符
DECODE (x, search _ value, result,default_value)	对 value 与 search_value 比较，相等返回 result,否则返回 default_value

1. TO_CHAR(x[,format])

将 x 转为字符串，其中 x 可以为数字或日期，参数 format 可以指定转换后的显示格式。具体格式参见表 10-5 和表 10-6。

表 10-5 时间值格式化参数

部分	参数	说明	例子
年份	YEAR	年份全部大写	TWENTY TEN
	Year	年份首字母大写	Twenty Ten
	YYYY	完整 4 位年	2010
	YY	年份后两位	10
月份	MONTH	月份全部大写	NOVEMBER
	Month	月份首字母大写	November
	MON	月份前三个字母大写	NOV
	Mon	月份前三个字母首字母大写	Nov
	MM	两位月份	11
周	WW	本年第几周	
	W	本月第几周	
日	DAY	周几全部大写	FRIDAY
	Day	周几首字母大写	Friday
	DY	前三个字母大写	FRI
	Dy	前三个字母首字母大写	Fri
	DDD	本年第几天	
	DD	本月第几天	
	D	本周第几天	
时	HH24	24 小时格式时间	23
	HH	12 小时格式时间	11
分	MI	两位分	
秒	SS	两位秒	

表 10-6 数字格式化参数

参数	说明	例子
9	返回指定位置数字,不足位前面补空格	select to_char(12345, '9999999') from dual; TO_CHAR(- - - - - - - 12345
0	返回指定位置数字,不足位数补 0	select to_char(12345, '0999999') from dual; TO_CHAR(- - - - - - - 0012345
.	在指定位置返回小数点	select to_char(5.678,'9.99') from dual TO_CH 5.68
,	在指定位置返回一个逗号	select to_char(5678,'99,999,999') from dual TO_CHAR(567 _____ 5,678
$	在数字开头返回一个美元符号	select to_char(5678,'$99,999') from dual TO_CHAR(_____ $5,678
L	在指定位置处返回本地货币符号,该符号来源于数据库参数 NLS_CURRENCY	select to_char(5678,'L99,999') from dual TO_CHAR(_____ ¥5,678

```
SQL> select to_char(sysdate,'yyyy-mm-dd') from dual;
TO_CHAR(SYSDATE,'YYYY-MM-DD')
-----------------------------
2013-08-27

SQL> select to_char(sysdate,'yyyy dd/mm') from dual;
TO_CHAR(SYSDATE,'YYYY DD/MM')
-----------------------------
2013 27/08

SQL> select to_char(1234567,'999,999,999') from dual;
TO_CHAR(1234567,'999,999,999')
------------------------------
   1,234,567
```

2. TO_DATE(x[,format])

将字符串 x 转为日期,其中参数 format 参见表 10-5,format 用于说明 x 的格式。

```
SQL> select to_date('2010-10-01','yyyy-mm-dd') from dual;
TO_DATE('2010-10-01','YYYY-MM-DD')
----------------------------------
2010-10-01

SQL> select to_date('2010,10/01','yyyy,mm/dd') from dual;
TO_DATE('2010,10/01','YYYY,MM/DD')
----------------------------------
2010-10-01
```

3. TO_NUMBER(x[,format])

将字符串 x 转为数字,其中参数 format 参见表 10-6,format 用于说明 x 的格式。

```
SQL> select to_number('12345') from dual;
TO_NUMBER('12345')
------------------
             12345

SQL> select to_number('12,345') from dual;
select to_number('12,345') from dual
                 *
ERROR at line 1:
ORA-01722: invalid number

SQL> select to_number('12,345','99,999') from dual;
TO_NUMBER('12,345','99,999')
----------------------------
                       12345
```

4. TRANSLATE(x,from_string,to_string)

在 x 中查找 from_string 中的字符,将其替换成 to_string 中对应的字符。

```
SQL> select translate('hello','abcdefghijklmnopqrstuvwxyz',
    'ABCDEFGHIJKLMNOPQRSTUVWXYZ') from dual;
TRANSLATE
---------
HELLO
```

5. DECODE(value,search_value,result,default_value)

对 value 与 search_value 比较,相等返回 result,否则返回 default_value。该函数允许在 SQL 中执行 if-then-else 逻辑。

```
SQL> select sname,gender from student;
SNAME                       GENDER
--------------------        ------
李颖                          1
张良                          2
孙小卉

SQL> select sname,decode(gender,1,'男',2,'女','第三性别') from student;
SNAME              DECODE(GENDER,1,'男',2,'女','第三性别')
----------         ------------------------------------
李颖                男
张良                女
孙小卉              第三性别
```

10.6 正则表达式函数

正则表达式函数可完成字符串的模式搜索，参数可以是字面值，也可以是表中的列值。表 10-7 列出了常用的正则表达式函数，表 10-8 列出了常用的正则表达式的元字符。

表 10-7 转换函数

函　　数	说　　明
REGEXP_LIKE(x,pattern[,match_option])	在 x 中查找 pattern 参数中定义的表达式
REGEXP_INSTR(x,pattern[,start[,occurrence]])	查找字符串位置
REGEXP_ REPLACE (x , pattern [, replace _ string [, start [,occurrence]]])	替换字符串
REGEXP_SUBSTR(x,pattern[,start[,occurrence]])	查找匹配的字符串

表 10-8 正则表达式元字符

元字符	说　　明
^	匹配字符串的开头位置
$	匹配字符串的末尾位置
*	匹配前面的字符 0 或多次
+	匹配前面的字符 1 或多次
?	匹配前面的字符 0 或 1 次
{n}	匹配前面的字符恰好 n 次
{n,m}	匹配前面的字符最少 n 次，最多 m 次
{n,}	匹配前面的字符最少 n 次
{,m}	匹配前面的字符最多 m 次
x|y	匹配 x 或 y
[abcd…. xyz]或[a—z]	匹配字符集和

续表

元字符	说明
[::]	指定字符类,常用字符类如下。 [:alphanum:]可以匹配字符 0～9、A～Z 和 a～z [:alpha:]可以匹配字符 A～Z 和 a～z [:blank:]可以匹配空格或 Tab 键 [:digit:]可以匹配数字 0～9
\d	数字字符
\D	非数字字符
\w	字母字符
\W	非字母字符
\s	空白字符
\S	非空白字符

1. REGEXP_LIKE(x,pattern[,match_option])

在 x 中查找 pattern 参数中定义的表达式。

match_option 含义如下: 'c' 区分大小写; 'T' 不区分大小写; 'n' 允许使用可以匹配任意字符的操作符; 'm' 将 x 作为一个包含多行的字符串。

```
SQL> select * from test;
    ID NAME
------- --------------------
     1  abc
     1  123

SQL> select * from test where regexp_like(name,'^\D+$');
    ID NAME
------- --------------------
     1  abc

SQL> select * from test where regexp_like(name,'^\d+$');
    ID NAME
------- --------------------
     1  123
```

2. REGEXP_INSTR(x,pattern[,start[,occurrence]])

查找字符串位置。

```
SQL> select regexp_instr('hi,how are you?','a\w+') from dual;

REGEXP_INSTR('HI,HOWAREYOU?','A\W+')
------------------------------------
                                   8
```

上面查询查找字符 a 开头的字符串在 'hi,how are you?' 中的位置。

3. REGEXP_REPLACE(x,pattern[,replace_string[,start[,occurence]]])
替换字符串。

```
SQL> select regexp_replace('hi,how are you?','h\w+','xxx') from dual;
REGEXP_REPLACE('HI,HOWAREYOU?','H\W+','XXX')
--------------------------------------------
xxx,xxx are you?
```

上面查询将 h 开头的字符串都替换成 xxx。

4. REGEXP_SUBSTR(x,pattern[,start[,occurrence]])
查找匹配的字符串。

```
SQL> select regexp_substr('hi,how are you?','a\w+') from dual;
REGEXP_SUBSTR('HI,HOWAREYOU?','A\W+')
-------------------------------------
are
```

小　　结

Oracle 的内置函数主要分为两类：单行函数和多行函数。多行函数主要指分组统计函数，这部分函数在第 9 章中已有介绍，本章主要讲解 Oracle 数据库的单行内置函数、字符函数、数字函数、日期函数、转换函数和正则表达式函数，并介绍了每个函数的语法和用法。

思　考　题

假设存在学生表 student，数据如下，完成后面的查询并写出语句。

```
stuno    stuname    gender    age    birthday
-------  --------   -------   ---    ---------
101      张x三x                      1990-8-25
201      x李四x      男               1991-7-1
301      王五六
```

1. 查询生日是当月月末的学生信息。
2. 查询生日是当月月初的学生信息。
3. 查询生日是当月 20 日的学生信息。
4. 根据当前日期及学生生日计算每名学生年龄并更新学生表 age 列。
5. 查询学生信息，如某项信息为空显示 'Unknown'。
6. 将学生姓名统一按照最长学生姓名从左补齐，补齐字符用 '♯'。
7. 按照出生年份分组统计学生人数，输出信息包括出生年和人数。
8. 按照出生年份和月份分组统计学生人数，输出信息包括出生年月和人数。

第 11 章　PL/SQL 基础

PL/SQL 是 Oracle 数据库提供的编程语言，可通过逻辑控制语句完成复杂功能的实现。本章将介绍 PL/SQL 编程的基本语法。

11.1　PL/SQL 简介

PL/SQL 是一种程序语言，叫做过程化 SQL（Procedural Language/SQL）。PL/SQL 为 Oracle 数据库提供了内置的、解释的和独立于操作系统的编程环境。PL/SQL 是 Oracle 数据库对 SQL 语句的扩展，在 SQL 语句的基础上增加了编程元素，它将数据库技术和过程化程序设计语言联系起来，是一种应用开发语言，可使用条件语句、循环语句进行逻辑控制，将 SQL 的数据操纵功能与过程化语言数据处理功能结合起来。PL/SQL 的使用，使 SQL 成为一种高级程序设计语言，使 SQL 程序执行效率更高。

使用 PL/SQL 可以编写具有很多高级功能的程序，虽然通过多个 SQL 语句可能也能实现同样的功能，但是相比而言，PL/SQL 具有更为明显的一些优点。

（1）能够使一组 SQL 语句的功能更具模块化程序特点。
（2）采用了过程性语言控制程序的结构。
（3）可以对程序中的错误进行自动处理，使程序能够在遇到错误的时候不会被中断。
（4）具有较好的可移植性，可以移植到另一个 Oracle 数据库中。
（5）集成在数据库中，调用更快。
（6）减少了网络的交互，有助于提高程序性能。

通过多条 SQL 语句实现功能时，每条语句都需要在客户端和数据库服务器端传递，占用了大量网络带宽，并且消耗时间较多，而在网络中传输的那些结果，往往都是中间结果，不是我们所关心的。而使用 PL/SQL 程序是因为程序代码存储在数据库中，程序的分析和执行完全在数据库内部进行，用户所需要做的就是在客户端发出调用 PL/SQL 的执行命令，数据库接收到执行命令后，在数据库内部完成整个 PL/SQL 程序的执行，并将最终的执行结果反馈给用户。在整个过程中，网络里只传输了很少的数据，减少了网络传输占用的时间，所以整体程序的执行性能会有明显的提高。

11.2　基本块结构

PL/SQL 是块状编程语言，程序单元可能是命名块或未命名块，未命名块通常称为匿名块，命名块是指可存储在数据库中的对象，包括过程、函数、包和触发器。下面是一个匿名块

程序结构:

```
[DECLARE]
    声明部分
BEGIN
    执行部分          //至少有一条语句.NULL;
[EXCEPTION]
    异常处理部分
END;
/
```

1. 声明部分

声明部分可选,以关键字 DECLARE 开始,包含类型定义和变量声明,以及局部函数或过程定义。

2. 执行部分

执行部分以关键字 BEGIN 开始、以关键字 END 结束,包含变量赋值、对象初始化、分支控制语句、循环控制语句、嵌套的 PL/SQL 匿名块以及对命名块的调用。执行块不能空,要求至少包含一条语句,否则在编译时语法分析报错,如没有语句需要执行,包含一条 NULL 语句,如下例:

```
BEGIN
    NULL;
END;
```

3. 异常处理部分

异常处理部分以关键字 EXCEPTION 开始、以关键字 END 结束,用于处理程序执行过程中出现的异常状况,作用类似于 Java 语言的 CATCH 语句。

4. /

斜线代表执行的意思,SQL * Plus 工具中需要使用。

命名块结构与匿名块结构稍有不同,命名块有头定义部分,头定义部分和执行块之间是声明部分。下面是一个函数程序结构:

```
FUNCTION function_name (参数列表) RETURN 返回数据类型 IS
    声明部分
BEGIN
    执行部分          //至少有一条语句.NULL;
[EXCEPTION]
    异常处理部分
END;
/
```

如果执行过程中需要将结果显示到控制台,需要启用 SQL * Plus 的环境变量 SERVEROUTPUT,具体设置语法如下:

SET SERVEROUTPUT ON [SIZE 缓冲区大小]

其中,SIZE 语句用于指定 PL/SQL 输出数据的缓冲区大小,如果输出数据量较大需要设置。

11.3 变量类型

PL/SQL 支持的变量类型主要有：标量变量和复合变量。标量变量只包含一个值，比如字符、日期或数字，而复合变量可包含多个值，比如记录类型、数组类型等。

11.3.1 标量数据类型

1. 数字类型

1) NUMBER

NUMBER 数据类型支持定点数和浮点数，用法与 SQL 中类似，声明 NUMBER 变量时，精度和小数位数都是可选值。

```
NUMBER [( precision, [scale] )]
var1 NUMBER;
var2 NUMBER(15);
var3 NUMBER(10,2);
```

2) BINARY_INTEGER

存储数据范围 −2 147 483 647～＋2 147 483 647，数字以 2 的补码二进制格式存储，因此名称中有 BINARY 字样，占用 4 个字节，只能在 PL/SQL 中使用。

```
var1 BINARY_INTEGER;
var2 BINARY_INTEGER : = 100;
```

3) PLS_INTEGER

PLS_INTEGER 数据类型和 BINARY_INTEGER 数据类型相同，表达整数，存储数据范围 −2 147 483 647～＋2 147 483 647，占用 4 个字节，只能在 PL/SQL 中使用。

```
var1 PLS_INTEGER;
var2 PLS_INTEGER : = 100;
```

4) BINARY_DOUBLE

双精度数字，用于科学计算，精度小于 NUMBER 类型。

```
var1 BINARY_DOUBLE : = 20d;        //赋值时用"d"声明是双精度数据
```

5) BINARY_FLOAT

单精度数字，用于科学计算，精度小于 NUMBER 类型。

```
var1 BINARY_DOUBLE : = 20f;        //赋值时用"f"声明是单精度数据
```

2. 字符类型

1) CHAR

可存储 32 767 个字节，多于 SQL 中 2000 字节。

```
var1 CHAR(100);                    //100 个字节
var2 CHAR;                         //默认 1 个字节
var3 CHAR(100 CHAR);               //100 个字符
```

2) VARCHAR2

可存储 32 767 个字节,多于 SQL 中 4000 字节。

```
DECLARE
    v1 CHAR(32767) : = 'a';
    v2 VARCHAR2(32767) : = 'a';
BEGIN
    dbms_output.put_line('v1: '|| length(v1));
    dbms_output.put_line('v2: '|| length(v2));
END;
/
```

输出结果:

```
v1: 32767
v2: 1
```

CHAR 类型按照定义长度分配内存,而 VARCHAR2 类型根据需要动态分配内存。如果一个字符需要两个字节表示,则最多可存 32 767 / 2 个字符,如果一个字符需要三个字节表示,则最多可存 32 767 / 3 个字符。

3) LONG

可存储 32 760 个字节,少于 SQL 中 2GB。LONG 类型只用于向后兼容,建议少用。

3. 日期类型

1) DATE

DATE 类型表示时间,包含"年月日时分秒"。

```
var1 DATE;
var2 DATE : = SYSDATE;
var3 DATE : = to_date('2013 - 08 - 01', 'yyyy - mm - dd');
```

2) TIMESTAMP

TIMESTAMP 类型表示时间,除包含"年月日时分秒"外,还可指定秒的小数位。

```
var1 TIMESTAMP;
var2 TIMESTAMP(3) : = SYSTIMESTAMP;
```

3) 时间间隔

时间间隔类型包括:INTERVAL DAY TO SECOND 和 INTERVAL YEAR TO MONTH。

```
DECLARE
    var1 INTERVAL DAY TO SEDOND;
    var2 INTERVAL YEAR TO MONTH;
BEGIN
    var1 : = '5 10:10';
    var2 : = '101 - 3';
    var2 : = INTERVAL '101' YEAR;
    var2 : = INTERVAL '2' MONTH;
END;
```

/

4. 布尔类型

PL/SQL 新增类型,取值 TRUE、FALSE 或 NULL。

```
var1 BOOLEAN;                              //默认 NULL
var2 BOOLEAN := TRUE;
var1 BOOLEAN := FALSE;
```

11.3.2 复合数据类型

复合数据类型主要包括：记录类型和集合类型。记录类型也称为结构体,通常包含相关元素的集合。集合是一个事物集合,事物可以是标量变量、记录等,集合类型包括数组、嵌套表和联合数组。

1. 记录类型

也称结构体,是包含一系列变量的复合变量,可以方便处理单行多列数据。访问记录类型变量数据时使用"."操作符。

记录类型定义：

```
TYPE type_name IS RECORD (
    element1 数据类型 [ 默认值 ],
    element2 数据类型 [ 默认值 ],
    …
);
```

记录中各元素的数据类型可以是 SQL 数据类型或是 PL/SQL 数据类型,另外可以通过关键字 DEFAULT 或 := 指定默认值,例如：

```
DECLARE
    TYPE record_type IS RECORD(
      id number default 1,
      name varchar2(20) := 'xxx',
      gender varchar2(10)
    );
    v1 record_type;
BEGIN
    v1.id := 100;
    v1.gender := 'male';
    dbms_output.put_line( v1.id || ',' || v1.name || ',' || v1.gender );
END;
/
输出结果为: 100 , xxx , male
```

记录数据类型变量也可以通过使用%ROWTYPE 隐式定义。

1) 基于表

```
rec tblstudent % ROWTYPE;
```

2）基于游标

```
CURSOR c IS select * from STUDENT;
rec c % ROWTYPE;
```

【例】 通过记录类型查询部门信息。

```
DECLARE
    TYPE dept_rec IS RECORD(
        name dept.name % TYPE,
        id dept.id % TYPE
    );
    CURSOR c IS select * from dept;
    rec dept_rec;        //或者 rec c % rowtype 或者 rec dept % rowtype
BEGIN
    OPEN c;
    LOOP
        FETCH c INTO rec;
            EXIT WHEN c % NOTFOUND;
            dbms_output.put_line(rec.id||','||rec.name);
    END LOOP;
    CLOSE c;
END;
/
```

【例】 INSERT 语句使用记录类型。

```
DECLARE
    rec dept % ROWTYPE;
BEGIN
    rec.id: = 100;
    rec.name: = '测试部门';
    INSERT INTO dept values rec;
END;
/
```

【例】 UPDATE 语句使用记录类型。

```
DECLARE
    rec dept % ROWTYPE;
BEGIN
    rec.id: = 100;
    rec.name: = '研发部门';
    UPDATE dept SET ROW = rec where id = 100;
end;
/
```

2. 数组

数组类型与传统的数组类似，元素类型相同，使用有序数字索引，大小固定。数组可以在 SQL 和 PL/SQL 中定义。

1) SQL 中定义数组类型

CREATE TYPE varray_name IS VARRAY(maximum_size) OF SQL 数据类型 [NOT NULL];

可以通过查看视图 USER_TYPES、USER_SOURCE 获得类型信息。

2) PL/SQL 中定义数组类型

TYPE varray_name IS VARRAY(maximum_size)
OF SQL 数据类型 | PL/SQL 数据类型 [NOT NULL];

数组类型有三种状态：定义的、初始化的和分配的。通过调用构造函数来初始化数组变量，如使用没有任何参数的构造函数初始化时不分配空间，需要存储空间通过 EXTEND 方法分配，如将数据放入数组的构造函数可分配空间，但不能超过数据最大可用空间 maximum_size。

【例】 在 PL/SQL 块中作为数据变量类型。

```
DECLARE
    TYPE nvar IS VARRAY(3) OF NUMBER;
    v1 nvar;
    v2 nvar := nvar();
    v3 nvar := nvar(1,2,3);
BEGIN
    v1 := nvar();
    v1.EXTEMD(1);
    v1(1):= 100;

    v2.EXTEND(2);
    v2(1):= 100;
    v2(2):= 200;

    v3(1):= 100;
    v3(2):= 200;
    v3(3):= 300;

    FOR i IN 1..v1.COUND LOOP
        dbms_output.put_line(v1(i));
    END LOOP;
    FOR i IN 1..v2.COUNT LOOP
        dbms_output.put_line(v2(i));
    END LOOP;

    FOR i IN 1..v3.COUNT LOOP
        dbms_output.put_line(v3(i));
    END LOOP;
END;
/
```

【例】 数组作为表列类型。

```
CREATE TYPE phone_var IS VARRAY(5) OF VARCHAR2(20);
CREATE TABLE emp(
    id number,
    name varchar2(20),
    phone phone_var
);
INSERT INTO emp(1, 'john', phone_var('110', '112', '119'));
UPDATE emp SET phone = phone_var('110', '112', '120') where id = 1;
```

3. 嵌套表

嵌套表数据类型与数组类型很类似,是具有 SQL 或 PL/SQL 数据类型的一维结构体。和数组最大的区别是它没有初始最大尺寸。嵌套表可以在 SQL 和 PL/SQL 中定义。

1) SQL 中定义嵌套表类型

CREATE TYPE varray_name IS TABLE OF SQL 数据类型 [NOT NULL];

可以通过查看视图 USER_TYPES、USER_SOURCE 获得类型信息。

2) PL/SQL 中定义嵌套表类型

TYPE varray_name IS TABLE OF SQL 数据类型 | PL/SQL 数据类型 [NOT NULL];

嵌套表也有三种状态:定义的、初始化的和分配的,使用与数组非常相似。

【例】 在 PL/SQL 块中作为数据变量类型。

```
DECLARE
    TYPE nvar IS TABLE OF NUMBER;
    v1 nvar;
BEGIN
    v1 : = nvar();
    v1.EXTEMD(2);
    v1(1): = 100;
    v1(2): = 200;

    FOR i IN 1..v1.COUND LOOP
        dbms_output.put_line(v1(i));
    END LOOP;
END;
/
```

4. 联合数组

联合数组的前身称作 PL/SQL 表、索引表,Oracle 10g 中改名为联合数组,是具有 SQL 或 PL/SQL 数据类型的一维结构体,是稀疏结构,而数组或嵌套表是密集结构。

联合数组类型只能在 PL/SQL 作用域中引用,当从匿名或命名块程序外部使用时一般定义在包规范中,不能用于 SQL 表列类型。支持数字和字符串索引,使用起来与 Java 编程语言中的哈希表很相近。

定义语法：

```
TYPE type_name IS TABLE OF { SQL 数据类型 | PL/SQL 数据类型 } [NOT NULL]
INDEX BY {PLS_INTEGER | BINARY_INTEGER | VARCHAR2(size) };
```

【例】 整型索引值，索引连续。

```
DECLARE
    TYPE t_index_table IS TABLE OF NUMBER BY BINARY_INTEGER;
    v1 t_index_table;
BEGIN
    v1 (-1) : = 1;
    v1 (0) : = null;
    v1 (1) : = 2;
    FOR i IN v1.FIRST.. v1.LAST LOOP
        dbms_output.put_line('it('|| i || ') = '|| v1 (i) );
    END LOOP;
END;
/
```

【例】 整型索引值，索引不连续。

```
DECLARE
    TYPE t_index_table IS TABLE OF NUMBER BY BINARY_INTEGER;
    v1 t_index_table;
    current BINARY_INTEGER;
BEGIN
    v1 (-1) : = 1;
    v1 (0) : = null;
    v1 (100) : = 2;
    FOR i IN 1.. v1.COUNT LOOP
       IF i = 1 THEN
            current : = v1.FIRST;
       ELSE
            current: = it.NEXT(current);
       END IF;
       dbms_output.put_line( 'v1('|| current || ') = '|| v1 (current) );
    END LOOP;
END;
/
```

【例】 字符串索引值。

```
DECLARE
    TYPE t_index_table IS TABLE OF NUMBER BY VARCHAR2(20);
    v1 t_index_table;
    current VARCHAR2(20);
BEGIN
    v1 ('monday') : = 1;
```

```
        v1 ('friday') : = null;
        v1 ('tuesday') : = 2;
    FOR i IN 1.. v1.COUNT LOOP
        IF i = 1 THEN
            current : = v1.FIRST;
        ELSE
            current: = it.NEXT(current);
        END IF;
        dbms_output.put_line( 'v1('|| current || ') = '|| v1 (current) );
    END LOOP;
END;
/
```

11.3.3 集合运算符

集合运算符一般用于集合之间的赋值或比较。

1. CARDINALITY 运算符

用于计算集合中的元素数,如果希望计算值不同的元素数需与 set 运算符配合使用。语法如下:

CARDINALITY (集合名称);

2. EMPTY 运算符

用于检查变量是否为 NULL。语法如下:

variable IS [NOT] EMPTY

3. MEMBER OF

用于判断左操作数是否是右操作数的集合中的成员。语法如下:

variable_name MEMBER OF collection_name

4. MULTISET EXCEPT

用于从第一个集合中删除另一个集合。语法如下:

Collection1 MULTISET EXCEPT collection2

5. MULTISET INTERSECT

计算两个集合交集。语法如下:

collection1 MULTISET INTERSECT Collection2

6. MULTISET UNION

计算并集,包含重复元素,如去除重复元素使用 distinct。语法如下:

collection1 MULTISET UNION [DISTINCT] collection2

7. SET

从集合中删除重复元素。语法如下:

SET(collection)

8. SUBMULTISET

判断集合 1 是否是集合 2 的子集。语法如下：

collection1 SUBMULTISET collection2

11.3.4 集合方法

1. COUNT

COUNT 是个函数，返回数组和嵌套表分配空间数目，返回联合数组元素数目。

2. DELETE

DELETE 是个过程，删除元素，可删除单个元素 DELETE(n)，也可删除范围内元素 DELETE(m,n)。

3. EXISTS

是个函数，判断指定下标元素是否存在于集合中。

EXISTS(i)

4. EXTEND

EXTEND 是个过程，分配存储空间。

EXTEND 分配 1 个空间。

EXTEND(n) 分配 n 个空间。

EXTEND(n,i) 分配 n 个空间，且将 i 指定下标对应元素内容复制到新分配的空间中。

5. FIRST

FIRST 是函数，返回集合最低下标值。

6. LAST

LAST 是函数，返回集合最高下标值。

7. LIMIT

LIMIT 是函数，返回集合可以使用的最高下标值。

8. NEXT(n)

NEXT(n)是函数，返回基于当前下标的集合的下一个下标值。比如：NEXT(current)。

9. PRIOR(n)

PRIOR(n)是函数，返回基于当前下标的集合的上一个下标值。比如：PRIOR(current)。

10. TRIM

TRIM 是过程，删除集合中的下标值。

TRIM：删除集合中最高元素。

TRIM(n)：从集合末端删除 n 个元素。

11.4 变量声明赋值

1. 变量声明语法

variable_name data_type [:= | DEFAULT] default_value;

1) variable_name

variable_name 是变量名,变量名以字母开头,可以包含字母、数字、$、_和♯号,不可以使用 Oracle 的保留字作为变量名。

2) data_type

变量的数据类型可以是 SQL 数据类型或是 PL/SQL 数据类型。数据类型可以直接指定,也可以通过%TYPE 指定表中列的类型,或是通过%ROWTYPE 指定表中行数据类型。

3) default_value

声明变量时可以指定初值,指定方式可以使用关键字 DEFAULT 或是赋值符号":="。如果未指定初值,默认为 NULL。

2. 变量赋值语法

left_operand : = right_operand;

其中左操作数 left_operand 必须是变量,而右操作数 right_operand 可以是值、变量或表达式。

【例】 直接指定类型定义变量。

```
DECLARE
    id1 NUMBER;                         //未赋初值,默认 NULL
    id2 NUMBER := 100;                  //赋初值 100
    n1 varchar2(20) DEFAULT 'hi';       //赋初值'hi'
BEGIN
    id1 := 100;
    id2 := id1;
    n1 := 'how are you';
END;
/
```

【例】 通过%TYPE 和%ROWTYPE 定义变量。

```
DECLARE
    id1 dept.deptno % TYPE;
    r1 dept % ROWTYPE;
BEGIN
    id1 := 1001;
    r1.deptno := 1002;
    r1.deptname := 'sales';
    INSERT INTO dept VALUES r1;
END;
/
```

变量赋值除了上述使用操作符":="外,也可以使用 SELECT INTO 语句,从表中获取列数据赋给变量。

【例】 通过 SELECT INTO 给单个变量赋值。

```
DECLARE
    n varchar2(20);
BEGIN
    SELECT name INTO n FROM dept WHERE deptid = 101;
    dbms_output.put_line(n);
END;
/
```

【例】 通过 SELECT INTO 给记录变量赋值。

```
DECLARE
    r dept % ROWTYPE;
BEGIN
    SELECT * INTO r FROM dept WHERE deptid = 101;
    dbms_output.put_line(r.deptid || ', '|| r.deptname);
END;
/
```

11.5 控制结构

PL/SQL 中控制结构主要是指条件语句和循环语句,其中条件语句主要是 IF 语句和 CASE 语句,循环语句中主要包括简单循环语句、WHILE 循环语句和 FOR 循环语句。

11.5.1 条件语句

1. IF 语句

IF 语句支持单分支语句和多分支语句。IF 语句以关键 IF 开始,以 END IF 关键字结束。IF 语句评估条件,条件满足时执行 THEN 后面的语句,如果不满足执行 ELSE 后的语句。

单分支语法如下:

```
IF condition THEN
    true_execution_block;
[ELSE
    false_execution_block; ]
END IF;
```

【例】 根据成绩输出。

```
DECLARE
    score NUMBER : = 85 ;
BEGIN
    IF score > 90 THEN
        dbms_output.put_line('优秀! ');
    ELSE
        dbms_output.put_line('一般');
    END IF;
END;
/
```

多分支语法如下：

```
IF condition THEN
    true_execution_block;
ELSIF condition THEN
    true_execution_block;
[ELSE
    false_execution_block; ]
END IF;
```

【例】 根据成绩输出。

```
DECLARE
    score NUMBER : = 85 ;
BEGIN
    IF score > 90 THEN
        dbms_output.put_line('优秀! ');
    ELSIF score > 80 THEN
        dbms_output.put_line('良好! ');
    ELSIF score > 80 THEN
        dbms_output.put_line('中! ');
    ELSIF score > 80 THEN
        dbms_output.put_line('及格! ');
    ELSE
        dbms_output.put_line('不及格');
    END IF;
END;
/
```

2. CASE 语句

CASE 语句在 9.3.4 节做过介绍，CASE 语句可以应用到 SQL 语句中，也可以在 PL/SQL 中使用。

【例】 根据性别数字输出结果，性别 1 输出"男"，2 输出"女"，其他输出"未知"。

```
DECLARE
    gender NUMBER : = 1;
BEGIN
    CASE gender
        WHEH 1 THEN
            dbms_output.put_line('男');
        WHEN 2 THEN
            dbms_output.put_line('女');
        ELSE
            dbms_output.put_line('未知');
    END CASE;
END;
/
```

【例】 根据成绩输出。

```
DECLARE
    score NUMBER : = 85 ;
BEGIN
    CASE
        WHEN score > 90 THEN
            dbms_output.put_line('优秀! ');
        WHEN score > 80 THEN
            dbms_output.put_line('良好! ');
        WHEN score > 80 THEN
            dbms_output.put_line('中! ');
        WHEN score > 80 THEN
            dbms_output.put_line('及格! ');
        ELSE
            dbms_output.put_line('不及格');
    END CASE;
END;
/
```

11.5.2 循环语句

1. FOR 循环

FOR 循环是编程语言中使用最多的循环语句，FOR 循环隐式管理开始和结束，也可通过显式的 CONTINUE 或 EXIT 语句跳过循环或强制从循环中退出。FOR 循环包括以下两种形式。

1) 数值 FOR 循环

在定义的数值范围内按顺序遍历，语法如下：

```
FOR index IN [REVERSE] starting_number..ending_number LOOP
    repeating_statement;
END LOOP;
```

其中，starting_number 和 ending_number 必须是整型，循环索引变量 index 作用域限于 FOR 循环，不管前面是否已经定义该变量，循环都将创建一个新的局部作用域变量。REVERSE 关键字可使遍历顺序颠倒。例：

```
DECLARE
index NUMBER : = 10;
BEGIN
    FOR index IN 1..10 loop
        dbms_output.put_line(i);
    END LOOP;
END;
/
```

2）游标 FOR 循环

在 SELECT 语句返回的行中循环。如果要遍历表或视图中的数据，使用游标 FOR 循环很方便。不能使用游标 FOR 循环迭代引用游标（REF CURSOR），要迭代引用游标只能使用简单循环和 WHILE 循环。游标 FOR 循环可以使用一个定义好的 CURSOR，也可以使用 SELECT 语句代替局部定义的 CURSOR。语法如下：

```
FOR index IN [cursor[(parameter list)] | subquery ] LOOP
    repeating_statement;
END LOOP;
```

此处的索引变量是游标返回的记录结构的引用，可以通过"."号将游标索引和列名进行组合读取数据。

【例】 操作定义好的游标变量。

```
DECLARE
    CURSOR c IS select * from test;
    rec test % ROWTYPE;
BEGIN
    FOR i IN c LOOP
        dbms_output.put_line(i.name);
    END LOOP;
END;
/
```

【例】 操作定义好的带参数的游标变量。

```
DECLARE
    CURSOR c(str varchar2) IS select * from test where name = str;
    rec test % ROWTYPE;
BEGIN
    FOR i IN c('hello') LOOP
        dbms_output.put_line(i.name);
    END LOOP;
END;
/
```

【例】 直接写 select 语句，隐式游标。

```
DECLARE
    rec test % ROWTYPE;
BEGIN
    FOR i IN (select * from test) LOOP
        dbms_output.put_line(i.name);
    END LOOP;
END;
/
```

2. 简单循环

简单循环要求管理循环索引和退出条件，语法如下：

```
LOOP
    repeating_statement;
    EXIT WHEN 退出条件;
END LOOP;
```

或是：

```
LOOP
    EXIT WHEN 退出条件;
    repeating_statement;
END LOOP;
```

其中"EXIT WHEN 退出条件"也可以写成如下形式：

```
IF 退出条件 THEN
    EXIT;
END IF;
```

【例】 输出 100 个数。

```
DECLARE
    i NUMBER: = 0;
BEGIN
    LOOP
        i: = i + 1;                         //循环索引
        EXIT WHEN i > 100;                  //退出条件
        dbms_output.put_line(i);            //重复语句
    END LOOP;
END;
/
```

【例】 操作一般游标。

```
DECLARE
    CURSOR c IS select * from test;
    rec test % ROWTYPE;
BEGIN
    OPEN c;
    LOOP
        FETCH c INTO rec;
        EXIT WHEN c % NOTFOUND;
        dbms_output.put_line(rec.name);
    END LOOP;
    CLOSE c;
END;
/
```

【例】 操作引用游标。

```
DECLARE
    TYPE ref_cursor IS REF CURSOR;
    c ref_cursor;
    rec test % ROWTYPE;
BEGIN
    OPEN c FOR select * from test;
    LOOP
        FETCH c INTO rec;
        EXIT WHEN c % NOTFOUND;
        dbms_output.put_line(rec.name);
    END LOOP;
    CLOSE c;
END;
/
```

3. while 循环

和简单循环一样,需要管理循环索引和退出条件,不同于简单循环的是,它检查进入循环条件。语法如下:

```
WHILE 循环条件 LOOP
    repeating_statement;
END LOOP;
```

【例】 输出 100 个数。

```
DECLARE
    i NUMBER: = 0;
BEGIN
    WHILE i < 100 LOOP
        i: = i + 1;                        //循环索引
        dbms_output.put_line(i);           //重复语句
    END LOOP;
END;
/
```

【例】 操作引用游标。

```
DECLARE
    TYPE ref_cursor IS REF CURSOR;
    c ref_cursor;
    rec test % ROWTYPE;
BEGIN
    OPEN c FOR select * from test;
    WHILE c % ISOPEN LOOP
        FETCH c INTO rec;
        EXIT WHEN c % NOTFOUND;
        dbms_output.put_line(rec.name);
    END LOOP;
    CLOSE c;
END;
/
```

11.6 游　　标

在数据库中，游标是一个十分重要的概念。游标提供了一种对从表中检索出的数据进行操作的灵活手段，就本质而言，游标实际上是一种能从包括多条数据记录的结果集中每次提取一条记录的机制。游标总是与一条 SQL 查询语句相关联，游标由结果集（可以是零条、一条或由相关的选择语句检索出的多条记录）和结果集中指向特定记录的游标位置组成。

游标有两种类型：隐式游标和显式游标。在声明块中定义的游标为显式游标，在执行块中的任何 DML 语句都是隐式游标，在使用 SELECT INTO 语句、SELECT BULK COLLECT INTO 语句或游标 FOR 循环中，也创建隐式游标。

11.6.1 隐式游标

PL/SQL 块中的每条 SQL 语句实际上都是隐式游标。每个 DML 语句执行后都可以使用%ROWCOUNT 属性知道语句改变的行数，游标属性参见表 11-1。

表 11-1　游标属性

属　　性	说　　明
%FOUND	当访问到一行数据时返回 TRUE
%ISOPEN	隐式游标总返回 FALSE，对打开的显式游标返回 TRUE
%NOTFOUND	不能访问行数据时，返回 TRUE
%ROWCOUNT	返回行数

【例】 单行隐式游标。

```
DECLARE
    d date;
BEGIN
    SELECT sysdate into d FROM dual;
    dbms_output.put_line(d);
    dbms_output.put_line(SQL%ROWCOUNT);
END;
/
```

【例】 多行隐式游标。

```
BEGIN
    UPDATE dept SET created = sysdate;
    IF SQL%FOUND THEN
        dbms_output.put_line(SQL%ROWCOUNT);
    END IF;
END;
/
```

11.6.2 显式游标

在声明块中定义的游标为显式游标,显式游标可以是静态的显式游标,也可以是动态的显式游标,静态显式游标 SELECT 语句不变,动态显式游标 SELECT 语句可根据参数执行不同查询。

FOR 循环中使用显式游标,可自动完成打开、获取和关闭,简单循环或是 WHILE 循环中需要使用 OPEN、FETCH 和 CLOSE 语句完成游标的打开、获取和关闭。游标的基本操作语句如下。

游标定义语法:

```
CURSOR cursor_name [(parameter_definition)] IS | AS subquery;
```

打开游标语法:

```
OPEN cursor_name[(parameter_value)];
```

获取数据语法:

```
FETCH cursor_name INTO variable1 [, variable2, ….];
```

关闭游标语法:

```
CLOSE cursor_name;
```

1. 静态显式游标

静态显式游标是不变的 SELECT 语句,下例通过显式游标读取表 dept 中数据,包括游标定义、打开、读取和关闭操作。

【例】 通过游标读取表 dept 数据。

```
DECLARE
    CURSOR c IS select * from dept;
    rec c%ROWTYPE;
BEGIN
    OPEN c;
    LOOP
        FETCH c INTO rec;
        EXIT WHEN c%NOTFOUND;
        Dbms_output.put_line( rec.deptid || ', ' || rec.deptname);
    END LOOP;
    CLOSE c;
END;
/
```

2. 动态显式游标

动态显式游标可以使用变量，也可以使用形参来改变 SELECT 语句行为，下面分别演示这两种情况。

【例】 使用变量的动态显式游标。

```
DECLARE
    did NUMBER := 101;
    CURSOR c IS select * from dept where deptid = did;
    rec c%ROWTYPE;
BEGIN
    OPEN c;
    LOOP
        FETCH c INTO rec;
        EXIT WHEN c%NOTFOUND;
        Dbms_output.put_line( rec.deptid || ',' || rec.deptname);
    END LOOP;
    CLOSE c;
END;
/
```

打开游标时，变量 did 的值被替换，变量可使用初值，也可以在执行块中重新赋值。有一点需要注意，局部变量名不能与表中列名相同。

【例】 使用参数的动态显式游标。

```
DECLARE
    CURSOR c(did number) IS select * from dept where deptid = did;
    rec c%ROWTYPE;
BEGIN
    OPEN c(101);
    LOOP
```

```
            FETCH c INTO rec;
            EXIT WHEN c % NOTFOUND;
            Dbms_output.put_line( rec.deptid || ',' || rec.deptname);
        END LOOP;
        CLOSE c;
    END;
    /
```

3. 系统引用游标

系统引用游标类似于 C 语言中的指针,指针指向哪块内存就可读哪块内存数据,系统引用游标可执行任何结果集。系统引用游标可以是弱类型的引用游标,也可以是强类型的引用游标,弱类型引用游标在定义游标时不指定返回类型,而强类型引用游标在定义时需指定返回类型。

系统引用游标定义语法:

```
TYPE ref_cursor IS REF CURSOR [RETURN catalog_object_name % ROWTYPE];
```

【例】 通过系统引用游标分别读取 dept 表和 emp 表。

```
DECLARE
    TYPE ref_cursor IS REF CURSOR;
    c ref_cursor;
    rec1 dept % ROWTYPE;
    rec2 emp % ROWTYPE;
BEGIN
    OPEN c FOR select * from dept;
    LOOP
        FETCH c INTO rec1;
        EXIT WHEN c % NOTFOUND;
        Dbms_output.put_line( rec1.deptid || ',' || rec1.deptname);
    END LOOP;
    CLOSE c;

    OPEN c FOR select * from emp;
    LOOP
        FETCH c INTO rec2;
        EXIT WHEN c % NOTFOUND;
        Dbms_output.put_line( rec2.empid || ',' || rec2.empname);
    END LOOP;
    CLOSE c;
END;
/
```

11.7 批语句

批语句可用于大数据量的选择、插入、更新以及删除,可减少 PL/SQL 和 SQL 引擎的交互次数,从而提高性能。批语句包括查询时使用的 BULK COLLECT INTO,以及在插入、更新和删除语句中使用的 FORALL。

批处理语句中可以使用批游标属性%BULK_EXCEPTIONS(i)和%BULK_ROWCOUNT(i),其中%BULK_EXCEPTIONS(i)记录某行在操作时是否遇到错误,%BULK_ROWCOUNT(i)记录某行是否修改成功,成功返回 1,失败返回 0。

11.7.1 BULK COLLECT INTO 语句

通常获取游标数据时使用 FETCH cursor_name INTO var1, var2 的形式,每次读取一行。自 Oracle 8i 起,Oracle 提供了 FETCH BULK COLLECT INTO 语句来批量获取游标中的数据,能显著提升数据读取效率。用 BULK COLLECT 子句取回查询结果至集合中,在返回到 PL/SQL 引擎之前,关键字 BULK COLLECT 告诉 SQL 引擎批量输出集合。该关键字能用于 SELECT INTO、FETCH INTO 和 RETURNING INTO 语句中。

1. 使用 SELECT BULK COLLECT INTO

【例】 直接将表数据批量读取到复合变量。

```
DECLARE
    TYPE n_type IS TABLE OF NUMBER;
    TYPE v_type IS TABLE OF VARCHAR2(20);
    ids n_type;
    names v_type;
BEGIN
    SELECT * BULK COLLECT INTO ids,names FROM dept;
    FOR i IN 1..ids.COUNT LOOP
        dbms_output.put_line(ids(i) || ',' || names(i) );
    END LOOP;
END;
/
```

首先定义嵌套表类型 n_type 和 v_type,然后定义两个集合变量 ids 和 names 用于存储部门表中的部门编号和部门名称,然后循环输出两个集合变量的内容。也可以定义一个记录类型的嵌套表,代码如下:

```
DECLARE
    TYPE rec_type IS RECORD(
        id NUMBER,
        name VARCHAR2(20)
    );
    TYPE collection IS TABLE OF rec_type;
    recs collection;
```

```
BEGIN
    SELECT * BULK COLLECT INTO recs FROM dept;
    FOR i IN 1..recs.COUNT LOOP
        dbms_output.put_line(recs(i).id || ',' || recs(i).name );
    END LOOP;
END;
/
```

2. 使用 FETCH BULK COLLECT INTO

【例】 批量获取游标数据。

```
DECLARE
    TYPE rec_type IS RECORD(
        id NUMBER,
        name VARCHAR2(20)
    );
    TYPE collection IS TABLE OF rec_type;
    recs collection;
    CURSOR c IS select * from dept;
BEGIN
    OPEN c;
    FETCH c BULK COLLECT INTO recs ;
    FOR i IN 1..recs.COUNT LOOP
        dbms_output.put_line(recs(i).id || ',' || recs(i).name );
    END LOOP;
    CLOSE c;
END;
/
```

在使用 FETCH 语句取数据时可使用 LIMIT 语句限制返回的最大行数,如下例。

```
DECLARE
    TYPE rec_type IS RECORD(
        id NUMBER,
        name VARCHAR2(20)
    );
    TYPE collection IS TABLE OF rec_type;
    recs collection;
    CURSOR c IS select * from dept;
BEGIN
    OPEN c;
    LOOP
        FETCH c BULK COLLECT INTO recs LIMIT 5;
        EXIT WHEN recs.COUNT = 0;
        FOR i IN 1..recs.COUNT LOOP
            dbms_output.put_line(recs(i).id || ',' || recs(i).name );
        END LOOP;
    END LOOP;
    CLOSE c;
END;
/
```

在循环的退出条件中使用了集合的 COUNT 方法，如果没取到数据返回 0 行，此处如果使用 c%NOTFOUND 返回不确定值。

3. 使用 RETURNING BULK COLLECT INTO 语句

使用 RETURNING INTO 子句取回 DML 结果至集合，可用于查看执行结果。

【例】 批量获取删除数据结果。

```
DECLARE
    TYPE list IS TABLE OF NUMBER;
    enum list;
BEGIN
    DELETE FROM DEPT RETURNING deptid BULK COLLECT INTO enum;
    FOR i IN 1..enum.COUNT LOOP
        dbms_output.put_line(enum(i));
    END LOOP;
END;
/
```

11.7.2 FORALL 语句

在发送语句到 SQL 引擎前，FORALL 语句告知 PL/SQL 引擎批量输入集合。尽管 FORALL 语句包含一个循环模式，它并不是一个 FOR 循环。其语法为：

```
FORALL index IN lower_bound..upper_bound
    sql_statement;
```

FORALL 和 FOR 相比，FORALL 效率要高很多，可用于 DML 操作。

1. 用于 INSERT 语句

批量插入要求 VALUES 子句使用集合，下例演示从 dept 表中读取数据然后批量插入到 dept1 表中。

【例】 批量插入数据。

```
DECLARE
    TYPE n_type IS TABLE OF NUMBER;
    TYPE v_type IS TABLE OF VARCHAR2(20);
    ids n_type;
    names v_type;
BEGIN
    SELECT * BULK COLLECT INTO ids,names FROM dept;
    FORALL i IN 1..ids.COUNT
        insert into dept1 values(ids(i), names(i) );
END;
/
```

2. 用于 UPDATE 语句

批量修改要求在 SET 和 WHERE 子句中使用标量集合，下例演示将 dept 表部门名称增加字符 xxx。

【例】 批量修改数据。

```
DECLARE
    TYPE n_type IS TABLE OF NUMBER;
    TYPE v_type IS TABLE OF VARCHAR2(20);
    ids n_type;
    names v_type;
BEGIN
    SELECT * BULK COLLECT INTO ids,names FROM dept;
    FOR i IN 1..ids.COUNT LOOP
        names(i) := names(i) || 'xxx';
    END LOOP;
    FORALL i IN 1..ids.COUNT
        Update dept set deptname = names(i) where deptid = ids(i);
    FOR i IN 1..ids.COUNT LOOP
        Dbms_output.put_line(SQL%BULK_ROWCOUNT(i));
    END LOOP;
END;
/
```

3. 用于 DELETE 语句

批量删除要求在 WHERE 子句中使用标量集合,下例演示删除 dept 表数据。

【例】 批量删除数据。

```
DECLARE
    TYPE n_type IS TABLE OF NUMBER;
    TYPE v_type IS TABLE OF VARCHAR2(20);
    ids n_type;
    names v_type;
BEGIN
    SELECT * BULK COLLECT INTO ids,names FROM dept;
    FORALL i IN 1..ids.COUNT
        Delete from dept where deptid = ids(i);
END;
/
```

11.8 异常处理

PL/SQL 通常在异常块中处理程序错误,PL/SQL 有两种错误:一种是编译错误,编译错误通常发生在编译阶段,另一种是运行错误,运行错误比较复杂,通常发生在执行块中,如果发生在声明块中将不能捕获处理。

系统异常包括系统预定义异常和用户定义异常。异常在异常块中捕获并处理,捕获语法如下:

```
EXCEPTION
    WHEN { predefined_exception | user_defined_exception | OTHERS } THEN
        exception_handling_statement;
```

WHEN 语句后指定要捕获的异常,可以是预定义异常 predefined_exception、用户定义异常 user_defined_exception,也可以是 OTHERS,使用 OTHERS 可以捕获所有异常。

异常相关函数见表 11-2。

表 11-2 异常函数

函 数	说 明	函 数	说 明
SQLCODE	返回错误码	SQLERRM	返回错误消息

11.8.1 预定义异常

Oracle 在 STANDARD 包中定义了一系列异常,有错误号和错误名称,见表 11-3。

表 11-3 常用预定义异常

异常	错误号	触发条件
COLLECTION_IS_NULL	ORA-06531	试图访问未初始化的嵌套表或数组
CURSOR_ALREADY_OPEN	ORA-06511	试图打开一个已经打开的游标
DUP_VAL_ON_INDEX	ORA-00001	试图向表列插入重复值
INVALID_NUMBER	ORA-01722	试图给数字赋非数字值
NO_DATA_FOUND	ORA-01403	试图访问不存在的数据
TOO_MANY_ROWS	ORA-01422	使用 SELECT INTO 返回多行
VALUE_ERROR	ORA-06502	将给一个变量赋给另一个不能容纳该变量时
ZERO_DIVIDE	ORA-01476	0 作除数

【例】 捕获除数为 0 的异常。

```
DECLARE
    i NUMBER : = 10;
BEGIN
    i: = i/0;
EXCEPTION
    WHEN ZERO_DIVIDE THEN
        dbms_output.put_line(sqlerrm);
END;
/
```

11.8.2 用户定义异常

1. 不绑定错误号

在 PL/SQL 中可像声明其他变量一样声明异常,异常需要根据程序逻辑引发,见下例。

```
DECLARE
    e EXCEPTION;
    age NUMBER : = 200;
BEGIN
    IF age > 150 THEN
        RAISE e;
    END IF;
EXCEPTION
    WHEN OTHERS THEN
        dbms_output.put_line(sqlcode ||', '|| sqlerrm );
END;
/
```

默认情况下，用户自定义异常错误码为 1，错误消息为 User-Defined Exception。

2. 绑定错误号

首先声明异常，然后声明编译器指令 PRAGMA，使用 EXCEPTION_INIT 指令将异常与 Oracle 错误码进行映射，EXCEPTION_INIT 指令的一个参数是用户定义的异常变量，第二个参数是错误码，这种异常不需要程序引发，运行时系统自动引发。下例将错误号 −01476（该错误码数据库已做定义，ZERO_DEVIDE）与用户定义异常绑定。

```
DECLARE
    myexp EXCEPTION;
    PRAGMA EXCEPTION_INIT(myexp, - 01476);
    i NUMBER : = 100;
BEGIN
    i : = i/0;
EXCEPTION
    WHEN myexp THEN
        dbms_output.put_line(sqlcode ||', '|| sqlerrm );
END;
/
```

3. 动态异常

动态的用户定义异常用于在程序执行过程中引发，不需定义，动态异常函数如下：

`RAISE_APPLICATION_ERROR(error_number,error_message);`

函数的第一个参数 error_number 指定错误号，范围 −20 000 ～ −20 999，第二个参数 error_message 指定错误消息，见下例。

```
DECLARE
    age NUMBER : = 200;
BEGIN
    IF age > 150 THEN
        RAISE_APPLICATION_ERROR( - 20001, '年龄太大,不能超过 150 ');
    END IF;
EXCEPTION
```

```
        WHEN OTHERS THEN
            dbms_output.put_line(sqlcode ||', '|| sqlerrm );
END;
/
```

11.9 动态 SQL

在 Oracle 数据库开发 PL/SQL 块中使用的 SQL 分为：静态 SQL 语句和动态 SQL 语句。所谓静态 SQL 指在 PL/SQL 块中使用的 SQL 语句在编译时是明确的，执行的是确定对象。而动态 SQL 是指在 PL/SQL 块编译时 SQL 语句是不确定的，如根据用户输入的参数的不同而执行不同的操作。编译程序对动态语句部分不进行处理，只在程序运行时动态地创建语句、对语句进行语法分析并执行该语句。

另外，DDL 语句及系统控制语句不能在 PL/SQL 中直接使用，要想实现在 PL/SQL 中使用 DDL 语句及系统控制语句，可以通过使用动态 SQL 来实现。

实现动态 SQL 有两种方式：使用 DBMS_SQL 包和本地动态 SQL（EXECUTE IMMEIDATE），本节只介绍本地动态 SQL。

本地动态 SQL 语法：

EXECUTE IMMEDIATE sql_statement
[INTO variable1[, variable 2…]]
[USING variable1[, variable 2…]];

【例】 执行 DDL 语句。

```
CREATE OR REPLACE PROCEDURE sp_ddl
(table_name VARCHAR2, field VARCHAR2, datatype VARCHAR2)
IS
    str_sql VARCHAR2(500);
BEGIN
    str_sql: = 'create table '||table_name||'('||field|| ' '||datatype ||')';
    EXECUTE IMMEDIATE str_sql;
END;
/
```

在执行 DDL 语句时，不允许使用占位符。

【例】 执行 DML 语句，实现登录验证函数。

```
CREATE OR REPLACE FUNCTION func_dml(name varchar2,pass varchar2)
RETURN BOOLEAN
IS
    num NUMBER : = 0;
    sqlstr VARCHAR2(300);
BEGIN
```

```
    sqlstr : = 'select count(name) from tbluser where name = :1 and pass = :2';
    EXECUTE IMMEDIATE sqlstr INTO num USING name,pass;
    IF num = 1 THEN
        RETURN TRUE;
    ELSE
        RETURN FALSE;
    END IF;
END;
/
```

【例】 执行 SELECT 语句,返回单行记录。

```
CREATE OR REPLACE PACKAGE ptype IS
    TYPE rec IS RECORD(id dept.deptid % TYPE, name dept.deptname % TYPE);
END;
/

CREATE OR REPLACE FUNCTION func_dml(id number)
RETURN ptype.rec
IS
    r ptype.rec;
    sqlstr VARCHAR2(300);
BEGIN
    sqlstr : = 'select * from dept where id = :1';
    EXECUTE IMMEDIATE sqlstr INTO r USING id;
    RETURN r;
END;
/

调用
DECLARE
    r ptype.rec;
BEGIN
    r: = func_dml(1);
    dbms_output.put_line(r.id);
    dbms_output.put_line(r.name);
END;
/
```

小　　结

本章讲解了 Oracle 数据库的 PL/SQL 编程基础知识,包括 PL/SQL 程序结构、变量类型定义、变量声明赋值、语句控制结构(分支语句 IF 和 CASE,循环语句 FOR 循环、简单循环和 WHILE 循环)、游标(静态游标、动态游标和引用游标)、异常处理,另外介绍了大数据量处理时批语句的运用,以及动态 SQL 语句的执行。

思 考 题

假设存在 t01 表和 t02 表，如下：

```
create table t01
        as select rownum n1,lpad('x',10) from dba_objects where rownum <= 10;
create table t02(id number,name varchar2(20),gender varchar2(20),age number);
```

1. 写出匿名 PL/SQL 程序的基本结构，并解释每部分的作用。
2. 使用简单循环语句编写一个 PL/SQL 程序完成如下功能：给定任意数 n，计算从 1 到 n 中能被 7 整除的整数个数。
3. 使用 while 循环语句编写一个 PL/SQL 程序完成如下功能：给定任意数 n，输出 1~n 之间的等比数列，如 1 2 4 8 16 32 …
4. 使用游标读取表 t01 所有数据并输出，要求使用简单循环语句。
5. 使用游标读取表 t01 所有数据并输出，要求使用 for 循环语句。
6. 编写一个 PL/SQL 程序，要求如下：
定义一个记录类型 T02REC，能够保存表 t02 行数据；
定义 T02REC 类型变量 r1，对 r1 进行赋值（要求值：101，张三，男，20），然后将 r1 插入到表 t02。
7. 编写一个 PL/SQL 程序，要求如下：
定义一个数组类型 TVAR，能够保存字符串；
定义一个 TVAR 类型的变量 str；
向变量 str 存入 10 个不同的字符串；
输出 str 中的所有数据。
8. 编写一个 PL/SQL 程序，要求如下：
定义一个记录类型 T02REC，能够保存表 T02 行数据；
定义一个嵌套表类型 ALLREC，能够保存 T02REC 类型数据；
定义一个游标 c 查询表 t02 所有数据；
定义一个 ALLREC 类型变量 v；
打开游标，批量读取游标 c 的数据到变量 v；
循环读取 v 中数据并输出。

第 12 章　PL/SQL 程序设计

通过使用 PL/SQL 可将功能独立的程序模块保存为数据库对象,也就是命名程序。本章主要介绍 Oracle 数据库的自定义函数、过程、包以及触发器的编写和使用。

12.1　函　　数

函数可从 SQL 语句或 PL/SQL 程序直接调用,每个函数必须有返回值。可在局部声明块中定义局部的函数,也可在数据库中定义。函数语法如下:

```
CREATE OR REPLACE FUNCTION function_name
[(
param1 [IN][OUT][NOCOPY] sql_datatype | plsql_datatype [ {: = | DEFAULT} default_value]
[, …]
)]
RETURN {sql_datatype | plsql_datatype}
[ AUTHID { DEFINER | CURRENT_USER } ]
[ RESULT_CACHE ]
{ IS | AS }
[ PRAGMA AUTONOMOUS_TRANSACTION; ]
    declaration_statements;
BEGIN
    execution_statements;
    RETURN variable;
[ EXCEPTION ]
    exception_handling_statements;
END;
/
```

属性说明:

1) 参数模式

IN:代表输入参数,传递值,不写参数类型默认为 IN。

OUT:代表输出参数,传递引用,如指定 NOCOPY 则直接应用变量地址,否则对副本操作。

IN OUT:代表参数既做输入又做输出。

注:默认情况下传递值,形参会复制一份实参的副本,然后在内部传递、修改等,发生异常,不会赋值给实参,控制权交还调用环境,而实参值不变,还是调用前的值。而使用了 NOCOPY 后,形参将获得一个指向实参的指针,然后在内部传递,赋值都直接修改实参,此

时如果异常发生,控制权交还调用环境,但是实参已经被修改了,无法还原成调用前的值。

【例】 两个数相加,设置两个输入型参数,通过函数返回值返回和。

```sql
SQL > CREATE OR REPLACE FUNCTION func_add(n1 NUMBER, n2 NUMBER)
    RETURN NUMBER
    IS
    BEGIN
        RETURN n1 + n2;
    END;
    /

SQL > SELECT func_add(10, 20) FROM DUAL;
FUNC_ADD(10,20)
---------------
             30
```

【例】 两个数相加,设置两个输入型参数,一个输出型参数,通过输出型参数返回和。

```sql
SQL > CREATE OR REPLACE FUNCTION func_add(n1 NUMBER, n2 NUMBER,
                                          n3 OUT NUMBER)
    RETURN NUMBER
    IS
    BEGIN
        n3 := n1 + n2;
        RETURN 0;
    END;
    /

SQL > DECLARE
        var1 NUMBER := 10;
        var2 NUMBER := 20;
        var3 NUMBER;
        res NUMBER;
    BEGIN
        res := func_add(var1,var2,var3);
        dbms_output.put_line('n1 + n2 = ' || var3);
    END;
    /
n1 + n2 = 30
```

2) 权限模型

AUTHID DEFINER | CURRENT_USER

DEFINER 为默认权限,定义者权限,任何有执行特权的人就像所有者那样操作。CURRENT_USER 为调用者权限,在执行函数时,使用调用者权限来操作模式对象。

【例】 定义者权限练习:用户定义一个函数 func_1,向表 t01 插入一条记录,授权给 B 用户执行该函数,用户 B 登录执行该函数。

```
C:\> SQLPLUS A/pass@remote
SQL> CREATE TABLE t01(id NUMBER);

SQL> CREATE OR REPLACE FUNCTION func_1
    RETURN NUMBER
    AUTHID DEFINER
    IS
    BEGIN
        insert into t01 values(100);
        return 0;
    END;
    /

SQL> GRANT EXECUTE ON func_1 to b;

C:\> SQLPLUS B/pass@remote
SQL> DECLARE
        res NUMBER;
    BEGIN
        res := a.func_1;
    END;
    /

C:\> SQLPLUS A/pass@remote
SQL> select * from t01;
    ID
    -------
    100
```

上例表明,用户 B 执行用户 A 的函数时,函数执行成功,A 的 t01 表插入一行数据,也就是函数执行时是以定义者 A 的权限执行的。

【例】 调用者权限练习:用户定义一个函数 func_1,向表 t01 插入一条记录,授权给 B 用户执行该函数,用户 B 登录执行该函数。

```
C:\> SQLPLUS A/pass@remote
SQL> CREATE TABLE t01(id NUMBER);

SQL> CREATE OR REPLACE FUNCTION func_1
    RETURN NUMBER
    AUTHID CURRENT_USER
    IS
    BEGIN
        insert into t01 values(100);
        return 0;
```

```
        END;
    /
SQL > GRANT EXECUTE ON func_1 to b;

C:\> SQLPLUS B/pass@remote
SQL > DECLARE
         res NUMBER;
      BEGIN
         res : = a.func_1;
      END;
    /
DECLARE
*
ERROR at line 1:
ORA - 00942: table or view does not exist
ORA - 06512: at "U01.FUNC_1", line 6
ORA - 06512: at line 4
```

上例表明用户 B 执行用户 A 定义的函数 func_1 时,函数中试图访问调用者 B 的表 t01,而 B 用户没有 t01 表,所以报错。

3) 结果缓存

RESULT_CACHE 是 Oracle 11g 中的新增内容,使函数执行结果可以缓存到 SGA 中,下次再执行同样的调用可直接从 SGA 中获得结果,提高性能。

【例】 两个数相加,将结果缓存。

```
SQL > CREATE OR REPLACE FUNCTION func_add(n1 NUMBER, n2 NUMBER)
      RETURN NUMBER
      RESULT_CACHE
      IS
      BEGIN
          RETURN n1 + n2;
      END;
    /
SQL > set autotrace trace exp
SQL > SELECT /* + result_cache */ func_add1(10, 10) FROM DUAL;
Execution Plan
----------------------------------------------------------
Plan hash value: 1388734953
----------------------------------------------------------
| Id | Operation        | Name                          | Rows |
----------------------------------------------------------
|  0 | SELECT STATEMENT |                               |  1   |
|  1 | RESULT CACHE     | 51n8smt03kkju9vy89cu7uw1cp    |      |
|  2 | FAST DUAL        |                               |  1   |
----------------------------------------------------------
```

4）自治事务

正常情况下函数被调用时，处于调用者的事务中，如果希望函数作为一个单独的程序单元来运行，不受调用者事务影响，可以使用 PRAGMA AUTONOMOUS_TRANSACTION 指明是自治事务，在执行部分可调用 COMMIT 或是 ROLLBACK，且数据操作不受调用者事务影响。

【例】 定义函数 func_2，向表 t01 插入数据，调用该函数，但调用后执行 ROLLBACK，观察插入数据情况。

```
C:\> SQLPLUS A/pass@remote
SQL> CREATE OR REPLACE FUNCTION func_2(n NUMBER)
    RETURN NUMBER
    IS
    BEGIN
        insert into t01 values(n);
        return 0;
    END;
    /

SQL> SELECT * FROM t01;
no rows selected.

SQL> DECLARE
        res NUMBER;
    BEGIN
        insert into t01 values(100);
        res:= func_2(200);
        rollback;
    END;
    /

SQL> SELECT * FROM t01;
no rows selected.
```

【例】 修改函数 func_2，指定事务自治。

```
C:\> SQLPLUS A/pass@remote
SQL> CREATE OR REPLACE FUNCTION func_2(n NUMBER)
    RETURN NUMBER
    IS
    PRAGMA AUTONOMOUS_TRANSACTION;
    BEGIN
        insert into t01 values(n);
        commit;
        return 0;
    END;
```

```
            /
SQL > SELECT * FROM t01;
no rows selected.

SQL > DECLARE
       res NUMBER;
     BEGIN
       insert into t01 values(100);
       res: = func_2(200);
       rollback;
     END;
     /

SQL > SELECT * FROM t01;
     ID
   -------
     200
```

在 SQL 语句中使用返回 SQL 数据类型的函数,返回 PL/SQL 数据类型的函数只能用于 PL/SQL 块中。不能在 SQL 查询中使用包含 DML 操作的存储函数,不过可在插入、更新和删除中调用执行 DML 操作的函数。如果一定需要在 SQL 查询中调用包含 DML 操作的函数需指明事务自治。

如果函数的参数只有 IN 类型参数,则函数可在 SQL 语句和 PL/SQL 程序中调用,如果函数有 OUT 类型或 IN OUT 类型参数,则只能在 PL/SQL 程序中调用,语法如下:

1) SQL 中调用

```
SELECT function_name(params) from DUAL;
```

2) PL/SQL 中调用

```
DECLARE
    target_variable_name type;
BEGIN
    target_variable_name: = function_name(params);
END;
```

12.2 过　　程

过程不能作为右操作数,也不能用于 SQL 语句。可在局部声明块中定义局部的过程,也可在数据库中定义。过程语法如下:

```
CREATE OR REPLACE PROCEDURE procedure_name
[ (
param1 [IN][OUT][NOCOPY] sql_datatype | plsql_datatype [ {: = | DEFAULT} default_value]
[, …]
) ]
[ AUTHID {DEFINER | CURRENT_USER } ]
```

```
{ IS | AS }
[ PRAGMA AUTONOMOUS_TRANSACTION; ]
    declaration_statements;
BEGIN
    execution_statements;
[ EXCEPTION ]
    exception_handling_statements;
END;
/
```

过程语法中的参数与函数语法中的参数意义相同，不再赘述，与函数的最大区别是没有返回值语句 RETURN。过程的执行可以在 SQL*Plus 中使用 EXECUTE 命令执行，或在 PL/SQL 程序中执行。语法如下：

```
SQL > EXECUTE proc_1;
```

或

```
SQL > BEGIN
        proc_1;
     END;
```

【例】 清除指定表数据。

```
SQL > CREATE OR REPLACE PROCEDURE proc_1(tablename varchar)
    IS
        sqltext varchar2(500);
    BEGIN
        sqltext : = 'TRUNCATE TABLE '||tablename;
        EXECUTE IMMEDIATE sqltext;
    END;
    /

SQL > EXECUTE proc_1('t01');
```

12.3 调用过程和函数

在 Oracle 11g 版本中，无论是在 SQL 中调用还是在 PL/SQL 程序中调用函数或过程时，都可以使用位置和命名表示法。

位置表示法意味着形参列表中的每个参数都需要提供值，且必须按参数顺序提供。命名表示法意味着可以通过参数名进行必要的参数传递，对于有默认值的参数可不提供值。除了可以单独使用位置表示法和命名表示法外，也可以混合使用。

假设存在下面函数完成三个数相加：

```
CREATE OR REPLACE FUNCTION func_add(n1 NUMBER,n2 NUMBER : = 10,n3 NUMBER)
RETURN NUMBER
IS
```

```
        RETURN n1 + n2 + n3;
    END;
```

下面分别演示位置表示法、命名表示法和混合表示法的调用方式。

1) 位置表示法

可通过下列代码调用函数：

```
SELECT func_add(1,2,3) total from dual;
    TOTAL
    -----
        6
```

2) 命名表示法

可通过下列代码调用函数：

```
SELECT func_add(n1 => 1, n3 => 3) total from dual;
    TOTAL
    -----
       14
```

参数 n2 没有提供值，则使用默认值 10。

3) 混合表示法

可通过下列代码调用函数：

```
SELECT func_add(1, n3 => 3) total from dual;
    TOTAL
    -----
       14
```

混合表示法有一个限制，所有位置表示法的参数必须先出现，且必须按照顺序给出，否则会引发异常。

12.4 包

包可以将多个函数或过程组合一起，由创建它们的用户模式所有。包的实现分为两个部分，一个是包规范，另一个是包体。在包规范中可以声明变量、定义类型、声明函数和过程。包体则负责实现声明的组件，组件实现中可调用包规范声明的变量和类型以及其他过程和函数。

在包中，函数或过程可以同名，但参数数目或类型不能完全相同，这称为"重载"，再有组件实现过程中调用的其他过程或函数，必须在被调用前声明，这称为"前向引用"。

12.4.1 包规范

包规范将包声明为一个黑盒，声明中发布了可用的函数和过程。编译包规范后，可用 SQL * Plus 命令 DESC 查看包内的函数和过程。包规范语法如下：

```
CREATE [OR REPLACE] PACKAGE package_name
```

```
[ AUTHID {DEFINER | CURRENT_USER } ]
{ IS | AS }
[ PRAGMA SERIALLY_RESUABLE; ]
    -- 变量声明
    -- 类型定义
    -- 异常定义
    -- 函数声明
        FUNCTION function_name [ (parameter_declaration ) ] RETURN datatype;
    -- 过程声明
        PROCEDURE rocedure_name [ (parameter_declaration ) ];
END [package_name];
```

参数说明如下：

(1) AUTHID {DEFINER | CURRENT_USER }

包级权限模型，包内的组件将继承包的操作模式。

(2) PRAGMA SERIALLY_RESUABLE

只能在包上下文中使用，如果使用则需在包规范和包主体中都使用它。当要共享变量时，这个很重要，因为它能保证每次调用它们时都处于起始状态。

默认情况下，包内的变量值在内存保留，如果声明 PRAGMA SERIALLY_RESUABLE，则包内变量每次都使用初始值。

【例】 定义变量，未指定 PRAGMA SERIALLY_RESUABLE，通过另一个存储过程访问。

```
SQL > CREATE OR REPLACE PACKAGE var IS
        num NUMBER: = 100;
    END;
    /

SQL > CREATE OR REPLACE PROCEDURE var_test IS
    BEGIN
        var.num: = var.num + 1;
        dbms_output.put_line(var.num);
    END;
    /
```

调用两次存储过程 var_test，显示结果 101 和 102，也就是变量值在内存保留。打开另一个客户端，调用存储过程 var_test，显示 101，也就是说变量值只在当前进程中重用。

【例】 定义变量，指定 PRAGMA SERIALLY_RESUABLE，通过另一个存储过程访问。

```
SQL > CREATE OR REPLACE PACKAGE var IS
    PRAGMA SERIALLY_REUSABLE;
        num NUMBER: = 100;
    END;
    /

SQL > CREATE OR REPLACE PROCEDURE var_test IS
```

PL/SQL 程序设计

```
BEGIN
    var.num: = var.num + 1;
    dbms_output.put_line(var.num);
END;
/
```

无论在当前客户端还是新客户端,调用过程 var_test,都显示 101,包被调用时创建包的副本,变量会被重新初始化。

12.4.2 包体

包体是包规范中声明的函数和过程的实现。包体原型与包规范原型非常相似,包体可以声明包规范中设置的一切。只是不能在包体内定义函数的 PRAGMA 指令,包体不能声明与包规范同名的变量。包体中也可以定义包规范中没有声明的函数和过程,但不能被包外的程序访问。包体语法如下:

```
CREATE [OR REPLACE] PACKAGE BODY package_name
{ IS | AS }
[PRAGMA SERIALLY_RESUABLE;]
    --变量声明
    --类型定义
    --异常定义
    --函数实现
    FUNCTION function_name[ (parameter_declaration ) ] RETURN datatype
    IS
        declaration_statements;
    BEGIN
        execution_statements;
        RETURN variable;
        [ EXCEPTION ]
            exception_handling_statements;
    END;
    --过程实现
    FUNCTION function_name[ (parameter_declaration ) ]
    IS
        declaration_statements;
    BEGIN
        execution_statements;
        [ EXCEPTION ]
            exception_handling_statements;
    END;
END;
```

【例】 创建包,包含函数和过程。

```
SQL > CREATE OR REPLACE PACKAGE p1 IS
        FUNCTION adding(a NUMBER, b NUMBER) RETURN NUMBER;
        PROCEDURE adding(a NUMBER, b NUMBER, res OUT NUMBER);
    END p1;
```

```
        /
SQL> CREATE OR REPLACE PACKAGE BODY p1 IS
        FUNCTION adding(a NUMBER,b NUMBER) RETURN NUMBER
        IS
        BEGIN
           return a + b;
        END;
        PROCEDURE adding(a NUMBER,b NUMBER, res OUT NUMBER)
        IS
        BEGIN
           res: = a + b;
        END;
    END p1;
    /
SQL> select p1.adding(10,10) from dual;
    P1.ADDING(10,10)
    --------------------
              20
```

12.5 触发器

数据库触发器是特殊的存储程序,可以调用 SQL 语句和 PL/SQL 函数与过程。但不能直接被调用,由数据库中触发事件来激发。

触发器的用途如下。

(1) 控制 DDL 语句的行为,如通过更改、创建或重命名对象。

(2) 控制 DML 语句的行为,如插入、更新和删除。

(3) 实施参照完整性、复杂业务规则和安全性策略。

(4) 在修改视图中的数据时控制和重定向 DML 语句。

(5) 通过创建透明日志来审核系统访问信息。

触发器分类如下。

1. 数据定义语言触发器

当创建、修改或删除数据库模式中的对象时会激发这些触发器,它们有助于控制和监控 DDL 语句的执行。

2. 数据操作语言触发器

当在表中执行插入、更新或删除数据的操作时激活这些触发器,可以分别用语句级或行级触发器类型对表上的所有修改或每行的修改激发一次触发器。可以在语句执行前检查数据或生成主键值。

3. 复合触发器

当在表中执行插入、更新或删除数据操作时,这些触发器同时充当语句级和行级触发器的角色。该触发器可以捕获 4 个计时点信息:激发语句前;激发语句中的每一行变化前;激发语句中的每一行变化后;激发语句后。

4. INSTEAD-OF 触发器

这些触发器可以停止 DML 语句执行,并重定向 DML 语句。常用于视图数据的操作。

5. 系统或数据库事件触发器

当数据库中的系统活动发生时触发这些触发器,如登录和退出事件。

12.5.1 数据定义触发器

1. 数据定义触发器语法

数据定义触发器,也就是 DDL 触发器,DDL 事件发生时触发。数据定义触发器语法如下:

```
CREATE [OR REPLACE] TRIGGER trigger_name
{BEFORE | AFTER | INSTEAD OF} ddl_event
ON { DATABASE | SCHEMA }
[WHEN (logical_expression)]
[DECLARE]
    declaration_statement;
BEGIN
    execution_statement;
END;
```

参数说明如下:

1) BEFORE | AFTER | INSTEAD OF

指定触发器主体内容执行时间,BEFORE 指在定义语句执行前,AFTER 指在定义语句执行后,INSTEAD OF 指替换执行指令。

2) ddl_event

常用 DDL 事件:alter、create、comment、drop、grant、rename、revoke、truncate 或 ddl(表示任意数据定义事件)。

3) ON { DATABASE | SCHEMA }

创建数据库级触发器需要 ADMINISTER DATABASE TRIGGER 系统权限,一般只有系统管理员拥有该权限。对于模式级触发器,为自己的模式创建触发器需要 CREATE TRIGGER 权限,如果是为其他模式创建触发器,需要 CREATE ANY TRIGGER 权限。

4) WHEN (logical_expression)

指定触发器执行条件,比如数据库级触发器可判断产生 DDL 事件的对象拥有者 when (ora_dict_obj_owner='TEST'),或判断特定的 DDL 事件 when (ora_sysevent='CREATE' or ora_sysevent='ALTER')等。

2. 常用事件属性函数

以下函数均是 SYS 用户函数的同义词,例如,ORA_SYSEVENT 是 SYS.SYSEVENT 的同义词,Oracle 建议用同义词。

1) ORA_CLIENT_IP_ADDRESS

返回客户端 IP 地址。

2) ORA_SQL_TXT

返回 SQL 文本,需要一个参数,类型为 ORA_NAME_LIST_T,是一个集合类型。

3) ORA_DICT_OBJ_NAME

返回 DDL 语句目标对象名。

4) ORA_DICT_OBJ_OWNER

返回 DDL 语句目标所有者。

5) ORA_DICT_OBJ_TYPE

返回 DDL 语句目标对象类型。

6) ORA_IS_ALTER_COLUMN

函数接受一个形参,它是一个列名。该函数返回 BOOLEAN 数据类型的真或假值。当列被更改时返回真,列未被更改时返回假。

例:

```
IF ora_is_alter_column('name') THEN
    ...
END IF;
```

7) ORA_IS_DROP_COLUMN

需要一个参数指定列名,该函数用于判断指定列是否被删除。

8) ORA_SYSEVENT

返回负责激发触发器的系统事件。

9) ORA_LOGIN_USER

函数不接受形参。该函数返回当前模式名,数据类型为 VARCHAR2。

用法如下:

```
BEGIN
    INSERT INTO logging_table VALUES (ora_login_user||' is the current user.');
END;
```

【例】 创建一个数据库级触发器,监控用户 TEST1 和 TEST2 的 DDL 动作。

```
C:\> SQLPLUS system/oracle@remote
SQL> CREATE USER test IDENTIFIED BY test;
SQL> GRANT connect,resource TO test;
SQL> GRANT ADMINISTER DATABASE TRIGGER TO test;
SQL> conn test/test@remote;
SQL> CREATE TABLE ddl(txt varchar2(1000));
SQL> CREATE OR REPLACE TRIGGER ddl_t BEFORE ddl ON DATABASE
    WHEN (ora_dict_obj_owner = 'TEST1' or ora_dict_obj_owner = 'TEST2')
    DECLARE
        sql_txt ORA_NAME_LIST_T;
        sql_stmt varchar2(1000);
    BEGIN
        FOR i IN 1..ORA_SQL_TXT(sql_txt) LOOP
            sql_stmt:= sql_stmt||' '||sql_txt(i);
        END LOOP;
        insert into ddl values(sql_stmt);
    END;
    /
```

【例】 创建一个模式级触发器,监控用户 TEST1 的 create 和 alter 动作。

```
SQL> CREATE OR REPLACE TRIGGER ddl_t BEFORE ddl ON TEST1.SCHEMA
    WHEN (ora_sysevent in('CREATE','ALTER'))
    DECLARE
        sql_txt ORA_NAME_LIST_T;
        sql_stmt varchar2(1000);
    BEGIN
        FOR i IN 1..ORA_SQL_TXT(sql_txt) LOOP
            sql_stmt:= sql_stmt||' '||sql_txt(i);
        END LOOP;
        insert into ddl values(sql_stmt);
    END;
    /
```

【例】 创建一个模式级触发器,监控用户 TEST1 的所有 DDL 动作。

```
SQL> CREATE OR REPLACE TRIGGER ddl_t BEFORE ddl ON TEST1.SCHEMA
    DECLARE
        sql_txt ORA_NAME_LIST_T;
        sql_stmt varchar2(1000);
    BEGIN
        FOR i IN 1..ORA_SQL_TXT(sql_txt) LOOP
            sql_stmt:= sql_stmt||' '||sql_txt(i);
        END LOOP;
        insert into ddl values(sql_stmt);
    END;
    /
```

12.5.2 数据操作触发器

数据操作触发器,也就是 DML 触发器,数据变化时触发。数据操作触发器语法如下:

```
CREATE [OR REPLACE] TRIGGER trigger_name
{BEFORE | AFTER }
{INSERT | UPDATE | UPDATE OF column1[,…] | DELETE }
ON table_name
[FOR EACH ROW]
[WHEN (logical_expression)]
[PRAGMA AUTONOMOUS_TRANSACTION;]
[DECLARE]
    declaration_statement;
BEGIN
    execution_statement;
END;
```

参数说明如下:

1) INSERT | UPDATE | UPDATE OF column1[,…] | DELETE

指定触发条件,可以是插入、修改、删除行数据时触发,也可以是修改某列时触发。

2) FOR EACH ROW

FOR EACH ROW 指定触发器为行级触发器,否则为语句级触发器。行级触发器有 FOR EACH ROW 子句、一个 WHEN 子句及 new 和 old 伪记录。

语句级触发器也称为表级触发器,语句级触发器不能使用 WHEN 子句,不能使用 new 和 old 伪记录。语句级触发器可以用来收集一些表操作的信息,也可对整个表的操作权限等进行控制。

【例】 记录操作表的用户和日期。

```sql
SQL> CREATE OR REPLACE TRIGGER dml_t1
    AFTER INSERT ON DEPT
    BEGIN
        insert into deptlog values(ORA_LOGIN_USER, SYSDATE);
    END;
    /
```

【例】 限制表修改操作事件 周一至周五,8:00~17:00。

```sql
SQL> CREATE OR REPLACE TRIGGER dml_t2
    BEFORE
    UPDATE OR INSERT OR DELETE
    ON DEPT
    BEGIN
        IF (to_char(sysdate,'dy','nls_date_language = English') in
            ('sat','sun') )
            OR to_char(sysdate, 'hh24')< '08'
            OR to_char(sysdate, 'hh24')> '17' then
                raise_application_error( -20500, '非工作时间禁止修改数据!');
        END IF;
    END;
    /
```

【例】 测试触发器嵌套。

```sql
SQL> CREATE SEQUENCE nested_id;
SQL> CREATE TABLE nested(id number, d date);
SQL> CREATE OR REPLACE TRIGGER dml_t3
    BEFORE INSERT ON nested
    BEGIN
        insert into nested values(nested_id.nextval, sysdate);
    END;
    /
```

行级触发器可以捕获新值和旧值,可用来审核数据合理性。可使用伪记录 new 和 old 访问修改前后的字段值。:new 和 :old 是触发器主体中的绑定变量,因为触发器声明和主体是两个独立的 PL/SQL 块。new 和 old 在行级触发器的作用域中声明,触发器声明是调用块,触发器主体是被调用块。在 WHEN 子句中可引用 new 或 old,在触发器主体内引用 new 和 old 需加":"。

【例】 替换部门名称中的空格。

```
SQL > CREATE OR REPLACE TRIGGER dml_t4
    BEFORE INSERT ON dept
    FOR EACH ROW
    WHEN(regexp_like(new.name,' '))
    BEGIN
        :new.name : = regexp_replace(:new.name,' ','-');
    END;
    /
```

【例】 生成主键,取表中部门编号 deptid 最大值加 1。

```
SQL > CREATE OR REPLACE TRIGGER dml_t5
    BEFORE INSERT ON dept
    FOR EACH ROW
    WHEN (new.id is null)
    BEGIN
        select nvl(max(deptid),0) + 1 into :new.deptid from dual;
    END;
    /
```

【例】 多触发事件的 DML 触发器,可通过条件谓词区分 DML 操作。

```
SQL > CREATE OR REPLACE TRIGGER dml_t6
    BEFORE INSERT OR UPDATE OR DELETE ON dept
    BEGIN
        IF ( to_char(sysdate,'hh24') not between '08'and '17')
          OR (to_char(sysdate,'d') in (1,7)) THEN
          CASE
              WHEN INSERTING THEN
                  raise_application_error( - 20001,'当前不能增加员工信息.');
              WHEN UPDATING THEN
                  raise_application_error( - 20001,'当前不能修改员工信息.');
              WHEN DELETING THEN
                  raise_application_error( - 20001,'当前不能删除员工信息.');
          END CASE;
        END IF;
    END;
    /
```

12.5.3 复合触发器

复合触发器可捕获 4 个计时点信息:①激发语句前;②激发语句中的每一行变化前;③激发语句中的每一行变化后;④激发语句后。在表中进行 DML 操作时,复合触发器既是语句级触发器也是行级触发器。

复合触发器不支持 EXCEPTION 语句块,但可以在每个计时点块中实现 EXCEPTION 语句块,复合触发器至少要实现一个计时点块。

复合触发器语法如下:

```
CREATE [OR REPLACE] TRIGGER trigger_name
FOR {INSERT | UPDATE | UPDATE OF column1[, …] | DELETE }
ON table_name
COMPOUND TRIGGER
    -- 激发语句前计时点
    [ BEFORE STATEMENT IS
        [declaration_statement;]
    BEGIN
        execution_statement;
    END BEFORE STATEMENT; ]
    -- 激发语句中的每一行发生变化前计时点
    [ BEFORE EACH ROW IS
        [declaration_statement;]
    BEGIN
        execution_statement;
    END BEFORE EACH ROW; ]
    -- 激发语句中的每一行发生变化后计时点
    [ AFTER EACH ROW IS
        [declaration_statement;]
    BEGIN
        execution_statement;
    END AFTER EACH ROW; ]
    -- 激发语句后计时点
    [ AFTER STATEMENT IS
        [declaration_statement;]
    BEGIN
        execution_statement;
    END AFTER STATEMENT; ]
END;
```

当想同时要语句级和行级触发器,复合触发器是最好的选择。

12.5.4 INSTEAD-OF 触发器

INSTEAD-OF 触发器用来截获 DML 语句,并用备用的程序代码替换。通常用于不可更新的视图数据操作。INSTEAD-OF 触发器语法如下:

```
CREATE [OR REPLACE] TRIGGER trigger_name
INSTEAD OF dml_statement ON object_name
[FOR EACH ROW]
[WHEN (logical_expression)]
[DECLARE]
    declaration_statement;
BEGIN
    execution_statement;
END;
```

【例】 更新视图数据(如果创建视图带有 with check option,不能创建 instead of 触发器)。

```
SQL> CREATE TABLE t1(id number, name varchar2(20));
SQL> CREATE TABLE t2(id number, age number);
SQL> CREATE VIEW tv AS select a.id, a.name, b.age from t1 a, t2 b where a.id = b.id;

SQL> CREATE OR REPLACE TRIGGER insteadof_t1
    INSTEAD OF INSERT ON tv
    FOR EACH ROW
    BEGIN
        insert into t1 values(:new.id, :new.name);
        insert into t2 values(:new.id, :new.age);
    END;
    /
```

12.5.5 系统事件触发器

系统事件触发器可用来审核服务器启动(STARTUP)、关闭(SHUTDOWN)、服务器错误(SERVERERROR),以及用户登录(LOGON)或退出(LOGOFF)活动。系统事件触发器语法如下:

```
CREATE [OR REPLACE] TRIGGER trigger_name
{BEFORE | AFTER} database_event
ON database
[DECLARE]
    declaration_statement;
BEGIN
    execution_statement;
END;
```

【例】 创建记录登录和退出信息的触发器。

```
SQL> CREATE TABLE logon(name varchar2(50), logontime date, logofftime date);

SQL> CREATE OR REPLACE TRIGGER logon
    AFTER LOGON ON database
    BEGIN
        insert into logon(name, logontime) values(ora_login_user, sysdate);
    END;
    /

SQL> CREATE OR REPLACE TRIGGER logoff
    BEFORE LOGOFF ON database
    BEGIN
        update logon set logofftime = sysdate where name = ora_login_user;
    END;
    /
```

12.5.6 触发器编译和启用

1. 重新编译

ALTER TRIGGER [schema.] trigger_name COMPILE;

作用：如果触发器调用了函数或过程，当函数或过程被删除或修改后，触发器被标记为 INVALID，必须重新编译。

2. 启用/禁用触发器

ALTER TRIGGER [schema.] trigger_name DISABLE | ENABLE;

作用：对表进行大量数据操作时，禁用触发器后可节省大量时间。

启用/禁用某个表的所有触发器：

ALTER TABLE table_name DISABLE | ENABLE ALL TRIGGERS;

12.6 相关视图

1. DBA_PROCEDURES

DBA_PROCEDURES 视图显示数据库中所有过程和函数信息，参见表 12-1。

表 12-1 DBA_PROCEDURES 主要列介绍

列 名	描 述	列 名	描 述
OWNER	所有者	OBJECT_TYPE	对象类型
OBJECT_NAME	包名	AUTHID	执行者权限
PROCEDURE_NAME	过程或函数名		

2. DBA_TRIGGERS

DBA_TRIGGERS 视图显示数据库中所有触发器信息，参见表 12-2。

表 12-2 DBA_TRIGGERS 主要列介绍

列 名	描 述
OWNER	所有者
TRIGGER_NAME	触发器名
TRIGGER_TYPE	触发器类型
TRIGGER_EVENT	触发事件
TABLE_OWNER	触发器基于表的所有者
BASE_OBJECT_TYPE	对象类型（TABLE、VIEW、SCHEMA、DATABASE）
TABLE_NAME	表或视图名
WHEN_CLAUSE	条件语句
STATUS	触发器状态
TRIGGER_BODY	触发器执行体

3. DBA_SOURCE

DBA_SOURCE 视图显示存储对象的文本,参见表 12-3。

表 12-3 DBA_SOURCE 主要列介绍

列 名	描 述	列 名	描 述
OWNER	所有者	TYPE	对象类型
NAME	对象名	TEXT	创建对象的 SQL 文本

小 结

本章讲解了 Oracle 数据库的命名程序,包括函数、过程、包以及触发器。函数和过程部分讲解了基本创建语法,以及调用方法;包部分讲解了包规范和包体,包中可以包含函数、过程、数据类型或变量;触发器部分讲解了 Oracle 数据库支持的数据定义触发器、数据操作触发器、INSTEAD-OF 触发器、系统事件触发器,以及新增的复合触发器。

思 考 题

1. 编写一个函数 FUNC1,要求如下:计算参数 n 的阶乘并返回。
2. 编写一个函数 FUNC2,要求如下:计算 1 到参数 n 之间的数字和并返回。
3. 编写一个函数 FUNC3,要求如下:计算参数 start 和 end 之间能被 5 整除的数字个数和这些数字的和,通过两个输出类型参数返回。
4. 编写一个过程 PROC1,要求如下:

给定两个输入参数,一个是当前日期 curdate,另一个是生日 birth,计算年龄并通过一个输出参数返回。

5. 编写一个包 MYPACK,要求如下:

定义函数 ADD,完成参数 n1 和参数 n2 相加;

定义函数 SUB,完成参数 n1 和参数 n2 相减;

定义过程 LOG,记录日志,将通过参数 msg 传入的消息记录到数据库表 logs,事务自治。

表 logs:create table logs(logdate date,logmsg varchar2(500));

6. 假设存在表 dept 和 emp,并基于这两个表创建视图 dept_emp_v。创建一个触发器 trig1,当向

视图插入数据时触发器被触发,将插入视图的数据分别插入 dept 和 emp 表中。

```
create table emp(empid number,empname varchar(20),deptid number);
create table dept(deptid number,deptname varchar(20));
create view dept_emp_v
    as select empid,empname,deptid,deptname from emp
    join dept on emp.deptid = dept.deptid;
```

7. 为表 stu 创建一个生成主键值的触发器 trig2，要求：创建一个插入数据前的触发器；为当前插入行生成 stuno 列的值，值为表中 stuno 列最大值加 1。

create table stu(stuno number,stuname varchar2(20),age number);

8. 为表 salary 创建一个触发器 trig3，要求：
自定义一个异常 exp；
判断当前日期和时间，如果是星期六或星期日或时间是 8 点之前或 17 点之后，通过抛出异常的方式终止 DML 操作；
捕获异常 exp，打印异常信息，并再次抛出异常。

create table salary(empno number,empname varchar2(20),salary number);

第四部分

Oracle数据库工具使用

第 13 章 工 具 使 用

"工欲善其事,必先利其器",Oracle 数据库提供了大量的工具,熟练使用这些工具可帮助提升工作效率。本章主要介绍三类工具,SQL * Plus、SQL * Loader 和 EXPDP/IMPDP。

13.1 SQL * Plus

SQL * Plus 是 Oracle 数据库默认安装的与 Oracle 数据库进行交互的客户端工具,历经 Oracle 的版本变化,是最稳定的客户端工具,也是最常用的工具,具有很强的功能。在 SQL * Plus 中,可以运行 SQL * Plus 命令、SQL 语句以及 PL/SQL 程序。

SQL 语句或 PL/SQL 程序执行后,都保存在一个被称为 SQL BUFFER 的内存区域中,并且只保存最近一条语句,如果语句语法错误,使用 SQL * Plus 的 EDIT 命令进行编辑再重新执行。除了 SQL 语句或 PL/SQL 程序,在 SQL * Plus 中执行的其他语句称为 SQL * Plus 命令,它们执行完后,不保存在 SQL BUFFER 的内存区域中,一般用来格式化输出结果。

SQL * Plus 主要作用如下。
(1) 数据库的维护,如启动、关闭等。
(2) 执行 SQL 语句。
(3) 执行 PL/SQL 程序。
(4) 执行 SQL 脚本。
(5) 导出数据、用户管理及权限维护等。

13.1.1 启动 SQL * Plus

Oracle 数据库软件安装后,在软件安装目录的子目录 BIN 中包含 SQL * Plus 的可执行文件 sqlplus。如果是在 Windows 操作系统安装,该执行文件的路径已经添加到系统环境变量 PATH 中,可通过系统命令 SET 或 ECHO %PATH% 查看。如果是 Linux 系统,则该执行文件路径需要用户手动添加到系统环境变量 PATH 中,可通过系统命令 ENV 或 ECHO $PATH 查看。

在 Windows 系统中查看环境变量 PATH,如图 13-1 所示。

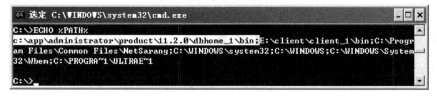

图 13-1 查看系统变量 PATH

在 Linux 系统中查看环境变量 PATH，如图 13-2 所示。

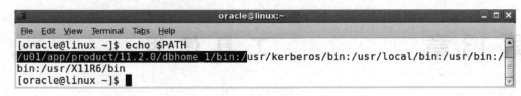

图 13-2　查看系统变量 PATH

打开一个命令行窗口，然后在命令提示符下输入"sqlplus"，然后回车，SQL * Plus 窗口显示如图 13-3 所示。

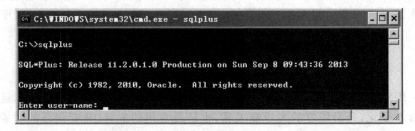

图 13-3　SQL * Plus 命令窗口

如果需要调整窗口显示文字字体和大小，可通过打开窗口的属性设置对话框进行调整，如图 13-4 所示。

图 13-4　窗口字体设置

13.1.2　SQL * Plus 环境变量设置

与 Oracle 数据库相关的系统环境变量设置将影响 SQL * Plus 的行为，常见的环境变量见表 13-1。

表 13-1 SQL * Plus 环境变量

环境变量	说 明
LD_LIBRARY_PATH	设置库文件路径 例：%ORACLE_HOME%/lib 或 $ORACLE_HOME/lib
LOCAL	Windows 环境中指定连接数据库的连接串（网络服务命名），Linux 中使用 TWO_TASK
NLS_LANG	指定语言、地区和字符集信息 例：american_america.zhs16gbk
ORACLE_HOME	指定 Oracle 数据库软件安装路径 例：c:\app\administrator\product\11.2.0\dbhome_1 或 /u01/app/product/11.2.0/dbhome_1
ORA_NLS10	指定 NLS 数据路径 例：%ORACLE_HOME%/nls/data 或 $ORACLE_HOME/nls/data
ORACLE_PATH	指定 SQL 脚本存放路径，否则只搜索当前目录，不适用于 Windows 系统
ORACLE_SID	指定 Oracle 实例名
PATH	系统可执行文件路径
SQLPATH	指定 SQL 脚本存放路径，适用于 Windows 系统
TNS_ADMIN	指定网络配置文件 tnsnames.ora 存放路径
TWO_TASK	指定连接数据库的连接串，Linux 中使用

【例】 环境变量 NLS_LANG 测试。

```
语言和地区设成美语和美国,登录后信息显示全英文
C:\> set NLS_LANG = american_america.zhs16gbk
C:\> sqlplus system/oracle@remote
SQL * Plus: Release 11.2.0.1.0 Production on Sun Sep 8 10:14:37 2013
Copyright (c) 1982, 2010, Oracle. All rights reserved.

Connected to:
Oracle Database 11g Enterprise Edition Release 11.2.0.1.0 - Production
With the Partitioning, OLAP, Data Mining and Real Application Testing options
SQL>

语言和地区设成汉语和中国,登录后部分信息显示中文
C:\> set NLS_LANG = simplified chinese_china.zhs16gbk
C:\> sqlplus system/oracle@linux_mydb
SQL * Plus: Release 11.2.0.1.0 Production on 星期日 9月 8 10:15:18 2013
Copyright (c) 1982, 2010, Oracle. All rights reserved.

连接到:
Oracle Database 11g Enterprise Edition Release 11.2.0.1.0 - Production
With the Partitioning, OLAP, Data Mining and Real Application Testing options
SQL>
```

【例】 环境变量 SQLPATH 测试。

```
假设存在 SQL 脚本文件 e:\test.sql
C:\> sqlplus system/oracle@remote
SQL>@test.sql
SP2-0310: unable to open file "test.sql"


C:\> set SQLPATH = e:\
C:\> sqlplus system/oracle@remote
SQL>@test.sql
SYSDATE
---------------
08-9月 -13
```

【例】 环境变量 LOCAL 测试。

```
通过 LOCAL 指定一个不存在的网络服务命名
C:\> set LOCAL = abc
C:\> sqlplus system/oracle
SQL*Plus: Release 11.2.0.1.0 Production on 星期日 9月 8 10:26:17 2013
Copyright (c) 1982, 2010, Oracle. All rights reserved.
ERROR:
ORA-12154: TNS:could not resolve the connect identifier specified

通过 LOCAL 指定一个存在的网络服务命名
C:\> set LOCAL = remote
C:\> sqlplus system/oracle
SQL*Plus: Release 11.2.0.1.0 Production on Sun Sep 8 10:14:37 2013
Copyright (c) 1982, 2010, Oracle. All rights reserved.


Connected to:
Oracle Database 11g Enterprise Edition Release 11.2.0.1.0 - Production
With the Partitioning, OLAP, Data Mining and Real Application Testing options
SQL>
```

13.1.3 SQL*Plus 配置

SQL*Plus 环境可通过全局配置文件 glogin.sql 或用户配置文件 login.sql 进行设置。

1. 全局配置文件

Linux 系统全局配置文件是 $ORACLE_HOME/sqlplus/admin/glogin.sql，Windows 系统全局配置文件是 %ORACLE_HOME%\sqlplus\admin\glogin.sql，全局设置对所有登录用户有效。

2. 用户配置文件

用户配置文件 login.sql 可以在当前目录中或 SQLPATH 指定的目录中。

【例】 修改 glogin.sql。

```
默认 glogin.sql
C:\> sqlplus system/oracle
SQL>

修改 glogin.sql,虚线中间是添加的内容
------------------------------------------------
SET SQLPROMPT _USER>
------------------------------------------------

再次登录查看提示信息
C:\> sqlplus system/oracle
SYSTEM>

修改 glogin.sql
------------------------------------------------
SET SQLPROMPT "_USER'@'_CONNECT_IDENTIFIER _DATE>"
------------------------------------------------

再次登录查看提示信息
C:\> sqlplus system/oracle
SYSTEM@linux_mydb 08-9月 -13>
```

【例】 修改 login.sql。

```
C:\> sqlplus system/oracle
SQL> select sysdate from dual;
SYSDATE
---------------
08-9月  -13

在当前目录编辑 login.sql
------------------------------------------------------------
ALTER SESSION SET nls_date_format = 'HH:MI:SS';
------------------------------------------------------------

C:\> sqlplus system/oracle
SQL> select sysdate from dual;
SYSDATE
----------
2013/09/08
```

13.1.4　SQL*Plus 连接数据库

使用 SQL*Plus 连接数据库语法如下：

SQLPLUS [[Options] [Logon|/NOLOG] [Start]]

语法属性说明：

1. Options 部分语法

-H[ELP] | -V[ERSION]

1) HELP 选项

可使用字母 H 或单词 HELP,显示 SQL*Plus 使用语法。

2) VERSION 选项

可使用字母 V 或单词 VERSION,显示 SQL*Plus 软件版本。

```
C:\> sqlplus -V
SQL*Plus: Release 11.2.0.1.0 Production
```

2. Logon 部分语法

username/password [@connect_identifier] [AS {SYSOPER|SYSDBA}]

1) username/password

指定连接数据库的用户名和口令,使用 SQL*Plus 工具连接数据库时,如果未指定用户名和口令,SQL*Plus 将提示用户输入。

```
C:\> sqlplus
SQL*Plus: Release 11.2.0.1.0 Production on Fri Oct 25 11:12:56 2013
Copyright (c) 1982, 2010, Oracle. All rights reserved.
Enter user-name: system
Enter password:

Connected to:
Oracle Database 11g Enterprise Edition Release 11.2.0.1.0 - Production
With the Partitioning, OLAP, Data Mining and Real Application Testing options
SQL>
```

2) @connect_identifier

通过网络连接远程数据库时要指定连接信息,@符号后可以直接写要连接数据库的 IP 地址、监听端口以及服务名,也可以写网络服务命名,建议使用后者。

```
C:\> sqlplus system/oracle@192.168.1.2:1521/mydb
SQL*Plus: Release 11.2.0.1.0 Production on Fri Oct 25 11:12:56 2013
Copyright (c) 1982, 2010, Oracle. All rights reserved.

Connected to:
Oracle Database 11g Enterprise Edition Release 11.2.0.1.0 - Production
With the Partitioning, OLAP, Data Mining and Real Application Testing options
SQL>
```

3) AS {SYSOPER|SYSDBA}

以超级用户连接数据库时需声明以什么身份进行连接,SYSDBA 和 SYSOPER 都是

Oracle 数据库的特殊权限,常用 SYSDBA。

```
C:\> sqlplus sys/oracle as sysdba
SQL*Plus: Release 11.2.0.1.0 Production on Fri Oct 25 11:12:56 2013
Copyright (c) 1982, 2010, Oracle. All rights reserved.

Connected to:
Oracle Database 11g Enterprise Edition Release 11.2.0.1.0 - Production
With the Partitioning, OLAP, Data Mining and Real Application Testing options
SQL>
```

4) NOLOG

不建立初始连接,之后通过命令 CONNECT 连接。

```
C:\> sqlplus /nolog
SQL*Plus: Release 11.2.0.1.0 Production on Fri Oct 25 11:12:56 2013
Copyright (c) 1982, 2010, Oracle. All rights reserved.

SQL> conn system/oracle
Connected.
```

3. Start 部分语法

@{url|file_name[.ext]} [arg ...]

指定连接数据库后要执行的脚本,可以通过@url 指定网络脚本,也可以通过@file_name 指定本地脚本,在进行环境准备时很有用。

```
编写脚本 e:\init.sql
--------------------------------------------------
conn system/oracle
select sysdate from dual;
quit
--------------------------------------------------

C:\> sqlplus /nolog @e:\init.sql
SQL*Plus: Release 11.2.0.1.0 Production on Fri Oct 25 11:18:10 2013
Copyright (c) 1982, 2010, Oracle. All rights reserved.
Connected.

SYSDATE
-----------
25-OCT-13

Disconnected from Oracle Database 11g Enterprise Edition Release 11.2.0.1.0 - Production
With the Partitioning, OLAP, Data Mining and Real Application Testing options
```

13.1.5　SQL*Plus 命令

常见的 SQL*Plus 命令见表 13-2。

表 13-2　SQL*Plus 命令

序号	命令	说明
1	@	执行指定的 SQL 脚本文件
2	@@	执行指定的 SQL 脚本文件
3	/	执行最近一次存储在 SQL Buffer 中的 SQL 语句或 PL/SQL 块程序
4	ACCEPT	接收一个输入值存储到指定变量中
5	ARCHIVE LOG	显示归档信息
6	CLEAR	清除设置信息
7	COLUMN	设置查询语句中列内容显示格式
8	CONNECT	连接数据库
9	DEFINE	定义变量
10	DEL	删除 SQL 缓冲区指定条目
11	DESCRIBE	显示指定对象的定义
12	DISCONNECT	断开与数据库的连接
13	EDIT	编辑指定文件或 SQL 缓冲区最近执行内容
14	EXECUTE	执行一个存储过程或执行一条 PL/SQL 语句
15	EXIT	退出 SQL*Plus
16	HELP	显示命令帮助
17	HOST	执行操作系统命令
18	LIST	显示 SQL 缓冲区指定条目
19	PASSWORD	修改口令
20	PRINT	显示所有变量或显示指定变量的值
21	PROMPT	显示提示字符信息
22	RECOVER	恢复数据库
23	SAVE	将 SQL 缓冲区的内容保存到操作系统文件
24	SET	设置环境变量
25	SHOW	显示信息
26	SHUTDOWN	关闭数据库
27	SPOOL	将查询结果写出到操作系统文件
28	START	执行指定的 SQL 脚本文件
29	STARTUP	启动数据库
30	STORE	存储系统环境变量
31	VARIABLE	声明变量

1. @符号

语法：@{url | file_name[.ext]} [arg…]

执行指定的 SQL 脚本文件，功能与@@和 START 相同。执行 SQL 脚本时可传递参数,参数在 SQL 脚本中通过 &1,&2,…来引用。

执行本地 SQL:

@test.sql

执行网络 SQL:

@HTTP://ip:port/xxx.sql var1 var2
@FTP://ip:port/xxx.sql var1 var2

【例】 执行本地 SQL。

```
test.sql 文件
conn &1/&2
show user


E:\> sqlplus /nolog @test.sql system oracle
SQL*Plus: Release 11.2.0.1.0 Production on Fri Oct 25 11:18:10 2013
Copyright (c) 1982, 2010, Oracle. All rights reserved.
Connected.
USER is "SYSTEM"
SQL>
```

2. @@符号

语法:@@{url | file_name[.ext]} [arg…]

执行指定的 SQL 脚本文件,功能与@和 START 相同,一般用在脚本嵌套的环境中,嵌套的脚本执行时在调用脚本的相同路径下查找。

【例】 连接数据库时执行脚本 s1.sql,在 s1.sql 中又调用 s2.sql。

```
s1.sql
conn &1/&2
@@s2.sql

s2.sql
create table t01(id number);
desc t01;

E:\> sqlplus /nolog @s1.sql system oracle
SQL*Plus: Release 11.2.0.1.0 Production on Fri Oct 25 11:18:10 2013
Copyright (c) 1982, 2010, Oracle. All rights reserved.
Connected.
Table created.
Name                    Null?            Type
----------------------  ---------------  -------------------
ID                                       NUMBER
```

3. / 符号

执行最近一次存储在 SQL Buffer 中的 SQL 语句或 PL/SQL 块程序。

```
SQL> select sysdate from dual;
SYSDATE
--------------
25 - OCT - 13

SQL>/
SYSDATE
--------------
25 - OCT - 13
```

4. ACCEPT 命令

语法：ACCEPT variable_name [NUMBER | CHAR | DATE] [FORMAT format] [PROMPT text]

接收一个输入值存储到指定变量中。FORMAT format 指定数据输入格式，PROMPT text 指定输入信息的提示字符。

```
SQL> accept var_id number prompt 'please input: '
please input: 1
SQL> select * from t01 where id = &var_id;
old    1: select * from t01 where id = &var_id
new    1: select * from t01 where id =         1
no rows selected.
SQL> define var_id
DEFINE VAR_ID          =         1 (NUMBER)
```

5. ARCHIVE LOG 命令

语法：ARCHIVE LOG LIST

显示归档信息。

```
SQL> archive log list;
Database log mode              Archive Mode
Automatic archival             Enabled
Archive destination            USE_DB_RECOVERY_FILE_DEST
Oldest online log sequence     17
Next log sequence to archive   19
Current log sequence           19
```

6. CLEAR 命令

语法：CLEAR [BUFFER | COLUMNS | SCREEN | SQL]

清除设置信息。CLEAR BUFFER 清除 SQL BUFFER，CLEAR COLUMNS 清除列的显示格式设置，CLEAR SCREEN 清除屏幕显示，CLEAR SQL 同 CLEAR BUFFER。

```
SQL> select * from t01;
        ID NAME
---------- ------------------------------
         1 abc

SQL> column name format a8
SQL> select * from t01;
        ID NAME
---------- --------
         1 abc

SQL> clear columns
columns cleared
SQL> select * from t01;
        ID NAME
---------- ------------------------------
         1 abc
```

7. COLUMN 命令

语法：COLUMN {column_name | alias | exp } [FORMAT format] [HEADING text]

设置查询语句中列内容显示格式。FORMAT format 指定显示格式，字符和日期型设置显示宽度 an，其中 a 代表字符，n 代表字符长度，数字型格式可参考函数 to_char 中数字格式。HEADING text 指定列显示标题名。

```
SQL> select * from t01;
        ID NAME
---------- ------------------------------
    123456 abc
SQL> col id format 999,999
SQL> col name format a20 heading 'Your Name is'
SQL> select * from t01;
      ID Your Name is
--------- --------------------
  123,456 abc
```

8. CONNECT 命令

语法：CONNECT username/password[@connect_identifier] [AS {SYSDBA | SYSOPER}]
以指定的用户名和口令、以特定的权限连接数据库，当前的连接断开。

```
C:\> sqlplus system/oracle
SQL*Plus: Release 11.2.0.1.0 Production on Mon Sep 9 09:29:00 2013
Copyright (c) 1982, 2010, Oracle. All rights reserved.
Connected to:
Oracle Database 11g Enterprise Edition Release 11.2.0.1.0 - Production
With the Partitioning, OLAP, Data Mining and Real Application Testing options
SQL> show user
```

```
USER is "SYSTEM"

SQL> connect sys/oracle as sysdba
Connected.
SQL> show user
USER is "SYS"
```

9. DEFINE 命令

语法：DEFINE [variable] | [variable = text]

定义一个变量，或列出已定义的变量。

```
SQL> define var1 = x12345
SQL> define var1
DEFINE VAR1            = "x12345" (CHAR)

SQL> define
DEFINE _DATE           = "09-SEP-13" (CHAR)
DEFINE _CONNECT_IDENTIFIER = "mysid" (CHAR)
DEFINE _USER           = "SYSTEM" (CHAR)
DEFINE _PRIVILEGE      = "" (CHAR)
DEFINE _SQLPLUS_RELEASE = "1102000100" (CHAR)
DEFINE _EDITOR         = "Notepad" (CHAR)
DEFINE _O_VERSION = "Oracle Database 11g Enterprise Edition Release 11.2.0.1.0 - Production
With the Partitioning, OLAP, Data Mining and Real Application Testing options" (
CHAR)
DEFINE _O_RELEASE      = "1102000100" (CHAR)
DEFINE VAR1            = "x12345" (CHAR)
```

10. DEL 命令

语法：DEL [n | n m | n * |n LAST | * | * n| * LAST|LAST]

删除 SQL 缓冲区指定条目。* 代表当前行，LAST 代表最后一行，n m 代表删除第 n 行到第 m 行，n * 代表删除第 n 行到当前行，n LAST 代表删除第 n 行到最后一行，* LAST 代表删除当前行到最后一行。

```
SQL> select * from t01;
        ID NAME
---------- --------------------
    123456 abc

SQL> list
  1* select * from t01
SQL> del 1
SQL> list
SP2-0223: No lines in SQL buffer.
```

11. DESCRIBE 命令

语法：DESCRIBE [schema .]object [@db_link]

显示指定对象的定义,比如表、视图、存储过程和函数。

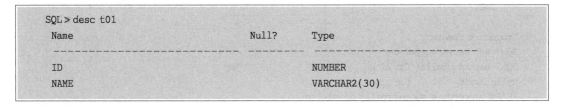

```
SQL> desc t01
 Name                             Null?    Type
 ------------------------------   -------- --------------------------
 ID                                        NUMBER
 NAME                                      VARCHAR2(30)
```

12. DISCONNECT 命令

语法:DESCONNECT

断开与数据库的连接,但不退出 SQL * Plus。

13. EDIT 命令

语法:EDIT [file_name]

指定 file_name 则编辑指定文件,否则编辑 SQL 缓冲区中最近执行的 SQL 语句。EDIT 命令执行时会自动打开一个文本编辑窗口。

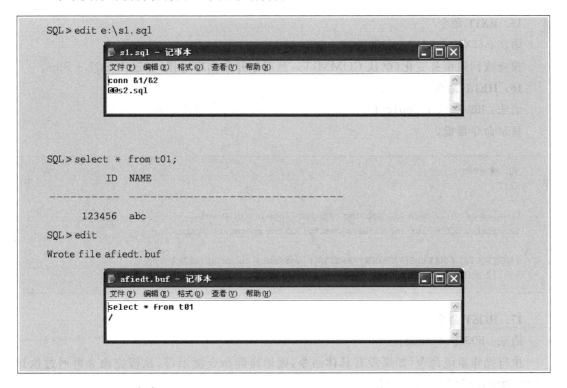

14. EXECUTE 命令

语法:EXECUTE statement

执行一个存储过程或执行一条 PL/SQL 语句。

```
执行存储过程
SQL> CREATE PROCEDURE hello(name varchar2)
    IS
    BEGIN
```

```
            dbms_output.put_line('Hello '|| name);
        END;
    /
Procedure created.
SQL> set serveroutput on
SQL> execute hello('David');
Hello David
PL/SQL procedure successfully completed.

执行 PL/SQL 语句
SQL> variable n number;
SQL> execute :n: = 100;
PL/SQL procedure successfully completed.
SQL> print n
    N
----------
    100
```

15. EXIT 命令

语法：{EXIT | QUIT } [COMMIT | ROLLBACK]

提交或回滚事务变化（默认 COMMIT），然后断开数据库连接，并退出 SQL * Plus。

16. HELP 命令

语法：HELP | ? [topic]

显示命令帮助。

```
SQL>? exit
EXIT
----
Commits or rolls back all pending changes, logs out of Oracle,
terminates SQL * Plus and returns control to the operating system.

{EXIT|QUIT} [SUCCESS|FAILURE|WARNING|n|variable|:BindVariable]
    [COMMIT|ROLLBACK]
```

17. HOST 命令

语法：HOST[command]

执行操作系统命令，如果没有具体命令，则切换到命令提示符，执行完命令再通过执行 EXIT 命令返回到 SQL 提示符。

```
SQL> host
Microsoft Windows XP [版本 5.1.2600]
(C) 版权所有 1985 – 2001 Microsoft Corp.
E:\> exit

SQL> host copy s1.sql s1.sql_bak
已复制         1 个文件.
```

18. LIST 命令

语法：LIST [n | n m | n * | n LAST | * | * n | * LAST | LAST]

显示 SQL 缓冲区指定条目。

19. PASSWORD 命令

语法：PASSWORD [username]

修改指定用户 username 的登录口令，或修改当前用户的登录口令。

```
SQL > password
Changing password for SYSTEM
Old password:
New password:
Retype new password:
Password changed

SQL > password u1
Changing password for u1
New password:
Retype new password:
Password changed
```

20. PRINT 命令

语法：PRINT [variable]

显示所有变量或显示指定变量的值。

```
SQL > print n
    N
-----------
    100
```

21. PROMPT 命令

语法：PROMPT [text]

显示提示字符信息。

22. RECOVER 命令

语法：RECOVER [AUTOMATIC] [FROM location]
    { DATABASE | TABLESPACE tablespace_name | DATAFILE filenumber}

恢复数据库，参考第 7 章内容。

23. SAVE 命令

语法：SAVE file_name [CREATE | REPLACE | APPEND]

将 SQL 缓冲区的内容保存到操作系统文件。可以创建新文件（CREATE），可以替换存在文件（REPLACE），或者追加到文件（APPEND）。

24. SET 命令

语法：SET system_variable value

设置环境变量。
SET ARRAY[SIZE] {15 | n}：设置每次取数据行数。
SET AUTO[COMMIT]{ON | OFF | IMM[EDIATE] | n}：设置自动提交模式，默认 OFF。
SET AUTOT[RACE] {ON | OFF | TRACE[ONLY]} [EXP[LAIN]] [STAT[ISTICS]]：设置跟踪。
SET HEA[DING] {ON | OFF}：设置列标题是否显示。
SET LIN[ESIZE] {80 | n}：设置行信息长度。
SET SERVEROUT[PUT] {ON | OFF} [SIZE {n | UNL[IMITED]}]：设置输出。
SET SQLP[ROMPT] {SQL>| text}：设置提示符。
SET TI[ME] {ON | OFF}：是否显示时间。
SET TIMI[NG] {ON | OFF}：是否显示计时信息。

25. SHOW 命令

语法：SHOW [system_variable | ALL] |
　　　　[ERRORS]
　　　　[PARAMETER parameter_name] |
　　　　[RECYCLEBIN] |
　　　　[RELEASE] |
　　　　[SGA] |
　　　　[USER]

显示系统环境变量、错误信息、系统参数、回收站、版本信息、SGA 信息和当前用户。

```
SQL > show timing
timing OFF

SQL > show user
USER is "SYSTEM"

SQL > show sga
Total System Global Area    41848422 bytes
Fixed Size                   1375004 bytes
Variable Size              289408228 bytes
Database Buffers           121634816 bytes
Redo Buffers                 6066176 bytes

SQL > show parameter memory_target
NAME                                 TYPE        VALUE
------------------------------------ ----------- ------------
memory_target                        big integer 400M
```

26. SHUTDOWN 命令

语法：SHUTDOWN [ABORT | IMMEDIATE | NORMAL | TRANSACTIONAL]
关闭数据库，参考第 2 章内容。

27. SPOOL 命令

语法：SPOOL file_name | OFF

将查询结果写出到操作系统文件。

28. START 命令

语法：START {url | file_name[.ext]} [arg…]

执行指定的 SQL 脚本文件，功能与@相同。

29. STARTUP 命令

语法：STARTUP [FORCE | RESTRICT | NOMOUNT | MOUNT | OPEN| PFILE = filename]

启动数据库，参考第 2 章内容。

30. STORE 命令

语法：STORE SET file_name

存储当前环境变量。

```
SQL> store set e:\env.txt
Created file e:\env.txt
```

31. VARIABLE 命令

语法：VARIABLE variable [type]

声明变量。

```
SQL> variable n number;
SQL> execute :n := 10;
```

13.2　SQL * Loader

SQL * Loader 是 Oracle 数据库的数据加载工具，通常用来将操作系统文本类型的数据文件导入到 Oracle 数据库中。SQL * Loader 可实现多个数据文件导入、多表导入，以及直接路径导入和并行导入，是实现数据加载的常用工具。

SQL * Loader 执行时需要一个或多个待加载的数据文件，需要一个控制文件，控制文件用来控制 SQL * Loader 的行为，另外在执行过程中会生成日志文件(log file)、记录错误数据的文件(bad file)以及记录废弃数据的文件(discard file)。具体输入输出信息参见图 13-5。

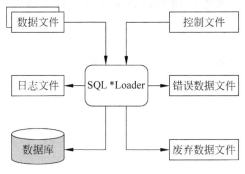

图 13-5　SQL * Loader 输入输出图

13.2.1 启动 SQL * Loader

Oracle 数据库软件安装后，在软件子目录 BIN 中包含 SQL * Loader 的可执行文件 sqlldr。如果是在 Windows 操作系统安装，该执行文件的路径已经添加到系统环境变量 PATH 中，可通过系统命令 SET 或 ECHO %PATH% 查看。如果是 Linux 系统，则该执行文件路径需要用户手动添加到系统环境变量 PATH 中，可通过系统命令 ENV 或 ECHO $PATH 查看。

打开命令行窗口，在命令提示符下输入命令"sqlldr"，显示命令 sqlldr 的用法，如图 13-6 所示。

图 13-6　命令 sqlldr 用法

SQL * Loader 命令的语法如下：

sqlldr keyword = value [,keyword = value, …]

其中，keyword 常用的有 userid、control 和 log，其中 userid 项指定连接数据库的用户名和口令，control 指定导入数据使用的控制文件，log 指定导入数据过程中生成的日志文件名。

假设存在表 T01，表中包含两列：ID 和 NAME，使用 SQL * Loader 将如下数据导入表中。
数据文件 d:\data.txt

101,张三
102,李四

控制文件 d:\load.ctl

```
------------------------------------------------------------
LOAD DATA
INFILE 'D:\DATA.txt'
APPEND
INTO TABLE T01
FIELDS TERMINATED BY ','
(
ID,
NAME
)
------------------------------------------------------------
执行导入:
D:\> sqlldr userid = u1/u1 control = load.ctl log = load.log
SQL * Loader: Release 11.2.0.1.0 - Production on 星期三 9月 11 13:58:50 2013
Copyright (c) 1982, 2009, Oracle and/or its affiliates. All rights reserved.

达到提交点 - 逻辑记录计数 1
达到提交点 - 逻辑记录计数 2
```

13.2.2 SQL * Loader 命令行参数

在命令提示符中执行没有参数的命令 sqlldr，则将显示所有 SQL * Loader 命令可用的参数，且部分参数有默认值，表 13-3 列出了常用的命令行参数。

表 13-3 SQL * Plus 命令行参数

序号	命令	说明
1	BAD	指定错误数据文件
2	CONTROL	指定控制文件
3	DATA	指定数据文件
4	DIRECT	指定直接路径导入，默认 FALSE
5	DISCARD	指定废弃数据记录文件
6	DISCARDMAX	指定最大允许废弃记录数
7	ERRORS	指定最大允许错误记录数
8	LOAD	指定最大加载数据行数
9	LOG	指定日志文件
10	PARFILE	指定参数文件
11	ROWS	指定每次提交的记录行数
12	SKIP	指定加载跳过的行数
13	USERID	指定连接数据库的用户名、口令

1. BAD(bad file)

通过 BAD 参数指定错误数据的记录文件，如果没有指定文件名，默认使用数据文件名称加扩展名.bad，例如数据文件为 data.txt，则默认错误数据记录文件名为 data.bad。在执行多文件导入时，在控制文件中可分别指定每个数据文件对应的错误数据记录文件。如果命令行参数和控制文件中都指定错误数据记录文件，则命令行参数指定的文件将被使用。

错误数据指格式错误的数据或向表中插入时报错的数据。

例：

```
SQLLDR u1/u1 control = load.ctl BAD = LOAD.BAD
```

2. CONTROL（control file）

通过 CONTROL 参数指定加载数据的控制文件，如果文件没有指定扩展名，默认是 .ctl。

例：

```
SQLLDR u1/u1 CONTROL = LOAD bad = load.bad
```

3. DATA（data file）

通过 DATA 参数指定加载的数据文件，通常加载的数据文件在控制文件中通过 INFILE 来指定，如果控制文件和命令参数都指定数据文件，则首先加载 DATA 参数指定的数据文件，且控制文件中指定的第一个文件将被忽略。

```
数据文件 d:\data1.txt
------------------------------------------------
101,张三
------------------------------------------------

数据文件 d:\data2.txt
------------------------------------------------
102,李四
------------------------------------------------

数据文件 d:\data3.txt
------------------------------------------------
103,王五
------------------------------------------------

控制文件 d:\load.ctl
------------------------------------------------
LOAD DATA
INFILE 'D:\DATA1.txt'
INFILE 'D:\DATA2.txt'
TRUNCATE
INTO TABLE T01
FIELDS TERMINATED BY ','
(
ID,
NAME
)
------------------------------------------------

执行导入：
D:\> sqlldr userid = u1/u1 control = load.ctl DATA = D:\DATA3.TXT log = load.log
SQL * Loader: Release 11.2.0.1.0 - Production on 星期三 9 月 11 13:58:50 2013
Copyright (c) 1982, 2009, Oracle and/or its affiliates. All rights reserved.
```

```
达到提交点 - 逻辑记录计数 1
达到提交点 - 逻辑记录计数 2
C:\> sqlplus u1/u1
SQL> select * from t01;
     ID  NAME
------  ----------------------
    103  王五
    102  李四
```

上面例子中共三个数据文件,控制文件中指定两个,命令行参数指定一个,结果显示控制文件中指定的第一个数据文件的内容没有导入。

4. DIRECT(data path)

通过 DIRECT 参数指定加载数据时使用直接路径加载,该参数默认是 FALSE。直接路径加载将要加载的数据直接进行数据块构建然后复制到数据库表,使用的是高水位线上的空间。

例:

SQLLDR u1/u1 control = load bad = load.bad DIRECT = TRUE

5. DISCARD(file name)

通过 DISCARD 参数指定废弃数据的记录文件,如果没有指定文件名,默认使用数据文件名称加扩展名.dsc,例如数据文件为 data.txt,则默认废弃数据记录文件名为 data.dsc。在执行多文件导入时,在控制文件中可分别指定每个数据文件对应的废弃数据记录文件。如果命令行参数和控制文件中都指定废弃数据记录文件,则命令行参数指定的文件将被使用。

废弃数据指不满足加载条件的数据,比如不满足控制文件中 WHEN 语句指定的条件。

例:

SQLLDR u1/u1 control = load.ctl bad = load.bad DISCARD = LOAD.DSC

6. DISCARDMAX(integer)

通过 DISCARDMAX 参数指定加载数据时最大的废弃记录数,如果废弃记录数超过该值,加载将中断。如果希望有任何一个废弃记录出现即中断加载,可设置 DISCARDMAX=0。

```
数据文件 d:\data.txt
----------------------------------------------------------
101,张三
102,李四
103,王五
----------------------------------------------------------

控制文件 d:\load.ctl
----------------------------------------------------------
LOAD DATA
INFILE 'D:\DATA1.txt'
TRUNCATE
INTO TABLE T01
WHEN ID <> '102'
FIELDS TERMINATED BY ','
(
```

```
ID,
NAME
)
------------------------------------------------------------
执行导入:
D:\> sqlldr userid = u1/u1 control = load.ctl log = load.log DISCARDMAX = 0
达到提交点 - 逻辑记录计数 2

D:\> sqlplus u1/u1
SQL> select * from t01;
      ID   NAME
------  --------------------
     101  张三
```

上例中加载数据到文件第二行时,数据不满足条件,由于设置 DISCARDMAX=0,导入立即中断,之前加载的数据提交,所以最后表中数据只有数据文件中的第一条。

7. ERRORS (error to allow)

通过 ERRORS 参数指定加载数据时最大的错误记录数,如果错误记录数超过该值,加载将中断。如果希望有任何一个错误记录出现即中断加载,可设置 ERRORS=0。

```
数据文件 d:\data.txt
------------------------------------------------------------
abc,张三
102,李四
103,王五
------------------------------------------------------------

控制文件 d:\load.ctl
------------------------------------------------------------
LOAD DATA
INFILE 'D:\DATA.txt'
TRUNCATE
INTO TABLE T01
FIELDS TERMINATED BY ','
(
ID,
NAME
)
------------------------------------------------------------

执行导入:
D:\> sqlldr userid = u1/u1 control = load.ctl log = load.log ERRORS = 0
达到提交点 - 逻辑记录计数 2

D:\> sqlplus u1/u1
SQL> select * from t01;
no rows selected.
```

上例中加载数据到文件第一行时,数据格式错误(第一列应为数字型,实际是字符串 abc),由于设置 ERRORS=0,导入立即中断,所以表中没有数据。

8. LOAD（number of records to load）

通过 LOAD 参数指定加载数据时最大的记录数,超过该值后剩余的数据行将被跳过。

例:

```
D:\< SQLLDR userid = u1/u1 control = load.ctl log = load.log load = 2
```

9. LOG（log file）

通过 LOG 参数指定加载数据时的日志文件,如果没有指定文件名,默认使用控制文件名称加扩展名.log,例如控制文件为 load.ctl,则日志文件为 load.log。

例:

```
D:\< SQLLDR userid = u1/u1 control = load.ctl LOG = LOAD.log
```

10. PARFILE（parameter file）

通过 PARFILE 参数指定命令行参数值。

```
参数文件 d:\load.par
USERID = u1/u1
CONTROL = load.ctl
LOG = load.log

D:\> sqlldr PARFILE = d:\load.par
```

11. ROWS（rows per commit）

通过 ROWS 参数指定每次提交的数据行数。

12. SKIP（records to skip）

通过 SKIP 参数指定数据文件开始的多少行被跳过不加载。

```
数据文件 d:\data.txt
--------------------------------------------------
101,张三
102,李四
--------------------------------------------------

控制文件 d:\load.ctl
--------------------------------------------------
LOAD DATA
INFILE 'D:\DATA.txt'
TRUNCATE
INTO TABLE T01
FIELDS TERMINATED BY ','
(
ID,
NAME
)
--------------------------------------------------
```

执行导入:
```
D:\> sqlldr userid = u1/u1 control = load.ctl skip = 1
达到提交点 - 逻辑记录计数 1

D:\> sqlplus u1/u1
SQL > select * from t01;
    ID NAME
  ----- -----
    102 李四
```

13. USERID（username/password）

通过 USERID 参数指定连接数据库的用户名和口令。

命令行参数除在执行命令时指定外，还可以在控制文件中通过 OPTIONS 语句来指定，该语句写在语句 LOAD DATA 前，语法如下：

OPTIONS = (keyword = value[,keyword = value])

```
控制文件 d:\load.ctl
-------------------------------------------------
OPTIONS = (rows = 2, skip = 1)
LOAD DATA
INFILE 'D:\DATA.txt'
TRUNCATE
INTO TABLE T01
FIELDS TERMINATED BY ','
(
ID,
NAME
)
```

13.2.3 控制文件格式说明

控制文件的基本格式如下：

```
[OPTIONS (keyword = value, … )]
LOAD DATA
INFILE 'datafile_name'
BADFILE 'badfile_name'
DISCARDFILE 'discardfile_name'
{ APPEND | REPLACE | INSERT | TRUNCATE }
INTO TABLE table_name
WHEN condition
FIELDS TERMINATED BY 'string'
TRAILING NULLCOLS
(
  column_list;
)
```

语句说明：

(1) OPTIONS：指定命令行参数。

(2) LOAD DATA：指定加载数据开始。

(3) INFILE：指定要加载的数据文件。

(4) BADFILE：指定加载数据文件对应的错误数据文件，可针对每个数据文件分别指定。

(5) DISCARDFILE：指定加载数据文件对应的废弃数据文件，可针对每个数据文件分别指定。

(6) APPEND | REPLACE | INSERT | TRUNCATE：指定加载数据方式，APPEND 是追加方式，REPLACE 方式先 DELETE 表中数据再 INSERT 数据，INSERT 方式直接插入数据但要求表必须为空，TRUNCATE 方式先 TRUNCATE 表中数据再 INSERT 数据。

(7) INTO TABLE：指定要插入数据的表。

(8) WHEN：指定数据过滤条件。

(9) FIELDS TERMINATED BY：指定数据行中列的分隔字符。

(10) TRAILING NULLCOLS：对于缺少数据的列赋空值。

13.2.4　指定加载数据文件

在语句 LOAD DATA 下使用 INFILE 语句指定要加载的数据文件，可加载一个文件，也可以加载多个数据文件，加载多个文件时写多个 INFILE 语句分别指定。

INFILE 语句语法如下：

```
INFILE input_filename | *
```

INFILE 使用方式有如下几种：

1. 指定星号 *

语法：INFILE *

指定星号的意思是数据文件包含在控制文件中，以语句 BEGINDATA 开始。

```
LOAD DATA
INFILE *
TRUNCATE
INTO TABLE t01
FIELDS TERMINATED BY ','
(id,
name)
BEGINDATA
101,张三
102,李四
```

2. 指定单个数据文件

语法：INFILE datafile_name BADFILE badfile_name DISCARD discardfile_name
　　　　　[DISCARDMAX n]

指定数据文件时可分别指定对应的错误数据文件和废弃数据文件，指定废弃数据文件时可指定最大废弃记录数。

```
LOAD DATA
INFILE 'd:\data1.txt' BADFILE 'd:\data1.bad' DISCARD 'd:\data1.dsc'
TRUNCATE
INTO TABLE t01
FIELDS TERMINATED BY ','
(id,
name)
```

3. 指定多个数据文件

控制文件中可通过多条 INFILE 语句指定多个数据文件。

```
LOAD DATA
INFILE 'd:\data1.txt' BADFILE 'd:\data1.bad' DISCARD 'd:\data1.dsc'
INFILE 'd:\data2.txt' BADFILE 'd:\data2.bad' DISCARD 'd:\data2.dsc'
TRUNCATE
INTO TABLE t01
FIELDS TERMINATED BY ','
(id,
name)
```

13.2.5 指定数据分隔方式

导入数据都有特定格式，有用特殊字符分隔的，也有数据定长通过位置指定的，具体使用如下。

1. 用字符分隔

导入数据时通常需要通过 FIELDS TERMINATED BY 语句指定数据文件中数据的分隔方式，常用语法如下：

```
TERMINATED BY {WHITESPACE | 'string' | X'hexstr'}
[OPTIONALLY ENCLOSED BY {'string' | X'hexstr'}]
```

其中，TERMINATED BY 语句指定行中数据分隔方式，WHITESPACE 指定使用空格或 Tab 分隔，或者直接指定分隔字符，比如 ','表示使用逗号分隔，或者指定分隔字符的 ASCII，比如指定逗号分隔也可写成 X'2c'。

数据除了使用字符分隔外，数据本身还有可能使用其他字符包围，比如数据行：101, '张三',20，这行数据首先使用了逗号进行数据分隔，另外中间的'张三'还使用了单引号进行包围，在插入数据库时不能包含单引号，这时声明数据分隔方式时使用如下语句：

```
TERMINATED BY ',' OPTIONALLY ENCLOSED BY ''''
```

```
LOAD DATA
INFILE *
TRUNCATE
INTO TABLE t01
TERMINATED BY ',' OPTIONALLY ENCLOSED BY ''''
```

```
(id,
name)
BEGINDATA
101, '张三'
102, '李四'
```

除了上面的用法外,还有如下用法:

```
TERMINATED BY {WHITESPACE | 'string' | X'hexstr'};
FIELDS ENCLOSED BY { 'string' | X'hexstr'};
TERMINATED BY {WHITESPACE | 'string' | X'hexstr'} ENCLOSED BY { 'string' | X'hexstr'};
```

如果只使用 ENCLOSED BY 语句,要求数据中两个分隔符间要有空格,否则当成一个分隔符。

2. 指定位置

数据分隔方式可以使用特定字符分隔,如果是定长数据也可以指定位置来获取数据,指定位置的语法为:

```
POSITION (start {:end | + length})
```

可以指定起止位置,比如:POSITION(1:5)代表从第 1 位到第 5 位。也可以指定起始位置和长度,比如:POSITION (1+5) 代表从第 1 位开始取 5 位长。

```
LOAD DATA
INFILE *
TRUNCATE
INTO TABLE t01
(id POSITION(1:4),
name POSITION(5:10)
BEGINDATA
101 张三
102 李四
```

13.2.6 指定条件

导入数据过程中可使用 WHEN 语句指定条件,将符合条件的数据导入表中。WHEN 语句语法如下:

```
WHEN condition
```

将年龄不等于 20 的员工导入。

```
LOAD DATA
INFILE *
TRUNCATE
INTO TABLE emp1
WHEN eage <> '20'
FIELDS TERMINATED BY ','
```

```
(
eid,
ename,
eage
)
BEGINDATA
101,李飞,20
102,赵刚,25
```

13.2.7 多表导入

单表数据导入在书中已有很多范例,下面讲述多表数据导入。多表操作时,同一行数据可记录多个表的列数据,导入时分别使用相应数据列,比如下面的数据:

```
101,李飞,20,D01,销售部
```

其中,前三列数据导入员工表 emp,后两列数据导入部门表 dept。

```
LOAD DATA
INFILE *
TRUNCATE
INTO TABLE emp
FIELDS TERMINATED BY ','
(
eid,
ename,
eage
)
INTO TABLE dept
FIELDS TERMINATED BY ','
(
did,
dname
)
BEGINDATA
101,李飞,20,D01,销售部
```

如果是同一行数据需要导入两个表,当导入第一个表时,行数据已经指到末尾,导入第二个表时需将数据重新定位从头开始,使用 POSITION(1)。比如下面将员工数据导入 emp1 和 emp2:

```
LOAD DATA
INFILE *
TRUNCATE
INTO TABLE emp1
FIELDS TERMINATED BY ','
(
eid,
```

```
ename,
eage
)
INTO TABLE emp2
FIELDS TERMINATED BY ','
(
eid POSITION(1),
ename,
eage
)
BEGINDATA
101,李飞,20
```

下面根据条件导入数据,将部门是"销售"的员工导入 emp1 表,将部门是"开发"的员工导入 emp2 表。

```
LOAD DATA
INFILE *
TRUNCATE
INTO TABLE emp1
WHEN dept = '销售'
FIELDS TERMINATED BY ','
(
eid,
ename,
eage,
dept
)
INTO TABLE emp2
WHEN dept = '开发'
FIELDS TERMINATED BY ','
(
eid POSITION(1),
ename,
eage,
dept
)
BEGINDATA
101,李飞,20,销售
102,赵刚,25,开发
```

13.2.8 指定列及数据类型

将数据导入表时,需要指定数据表列和数据文件中数据列的对应关系,以及数据列的类型和格式。

1. 指定 FILLER 数据列

如果数据文件中数据列较数据库表中列少,只要正确指定数据列和数据表列对应关系即可,如果数据文件中数据列多于数据库表中列,则需指定 FILLER 语句。FILLER 语句指定的数据列不插入到数据库表中。

```
LOAD DATA
INFILE *
TRUNCATE
INTO TABLE emp
FIELDS TERMINATED BY ','
(
dname filler,
eid,
ename
)
BEGINDATA
销售,101,李飞
开发,102,赵刚
```

2. 数据列类型和格式

在指定数据库表列和数据文件数据列对应关系时,语法如下:

```
table_columnname datafile_column_specification [, … ]
```

其中,table_columnname 指定数据库表列名,datafile_column_specification 指定数据列的类型和格式。在 SQL * Loader 读取数据时,如果未指定类型,默认都是 CHAR 类型,在插入数据库表时可隐式转换为数据库表数据类型,但如果是日期,且与系统默认格式不同时将报错,这时需要指定类型和格式,语法如下:

```
DATE 'date_format'
```

```
DATE 'date_format'
LOAD DATA
INFILE *
TRUNCATE
INTO TABLE emp
FIELDS TERMINATED BY ','
(
eid,
ename,
birth DATE 'yyyy/mm/dd'
)
BEGINDATA
101,李飞,1990/10/1
102,赵刚,1991/12/15
```

数据的分隔方式除了在 INTO TABLE 语句下进行设置外,也可以在每个数据列声明中指定,每个列的分隔方式都可以不同。

```
LOAD DATA
INFILE *
TRUNCATE
INTO TABLE emp
FIELDS TERMINATED BY ',' 
(
eid TERMINATED BY ',',
ename TERMINATED BY WHITESPACE,
birth DATE 'yyyy/mm/dd' TERMINATED BY '\n'
)
BEGINDATA
101,李飞 1990/10/1
102,赵刚 1991/12/15
```

3. 使用 NULLIF 和 DEFAULTIF 语句

NULLIF 指当条件成立时,将当前列设置 NULL；DEFAULTIF 指当条件成立时,将当前列设置 NULL 或 0,数字型设成 0,其他为 NULL。

```
LOAD DATA
INFILE *
TRUNCATE
INTO TABLE t01
FIELDS TERMINATED BY ','
(id INTEGER EXTERNAL DEFAULTIF id = '101',
name NULLIF name = 'b')
BEGINDATA
101,a
102,b

SQL> select * from t01;
      ID  NAME
   ------- ---------------
        0  a
      102
```

13.3　EXPDP

EXPDP 是 Oracle 数据库提供的数据泵导出实用程序,用于在 Oracle 数据库之间进行数据传输。常用于数据库的逻辑备份和不同平台间的数据移植。使用数据泵技术进行数据导出时,涉及以下三个部分。

(1) 命令行程序 expdp：通过命令行程序启动数据导出任务。

(2) 包 DBMS_DATAPUMP：命令行程序调用该包来完成数据导出。

(3) 包 DBMS_METADATA：用于数据库元数据的移动。

本节中主要介绍命令行程序 EXPDP 的使用方法。

13.3.1 启动 EXPDP

Oracle 数据库软件安装后,在软件子目录 BIN 中包含 EXPDP 的可执行文件 expdp。如果是在 Windows 操作系统安装,该执行文件的路径已经添加到系统环境变量 PATH 中,可通过系统命令 SET 或 ECHO %PATH% 查看。如果是 Linux 系统,则该执行文件路径需要用户手动添加到系统环境变量 PATH 中,可通过系统命令 ENV 或 ECHO $PATH 查看。

打开命令行窗口,在命令提示符下输入命令"expdp - help",显示命令 expdp 的用法,如图 13-7 所示。

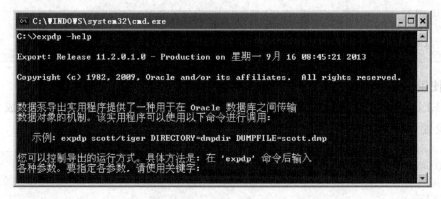

图 13-7 命令 expdp 用法

EXPDP 命令的语法如下:

expdp username/password[@connect_identifier] keyword=value [,keyword=value,…]

其中,username/password 指定执行导出任务的用户名和口令,connect_identifier 指定连接远程数据库的网络服务名,keyword 指定命令参数,常用的有 directory、dumpfile 等,directory 指定一个目录对象,dumpfile 指定导出文件名称。下例演示用户 u1 导出表 emp。

```
SQL> CREATE DIRECTORY mydir AS 'd:\exp';
目录已创建.
SQL> GRANT READ,WRITE ON DIRECTORY mydir TO u1;
授权成功.

C:\> expdp u1/u1 directory=mydir dumpfile=u1.dmp tables=emp
Export: Release 11.2.0.1.0 - Production on 星期一 9月 16 08:55:34 2013
Copyright (c) 1982, 2009, Oracle and/or its affiliates. All rights reserved.

连接到: Oracle Database 11g Enterprise Edition Release 11.2.0.1.0 - Production
With the Partitioning, OLAP, Data Mining and Real Application Testing options
启动 "U1"."SYS_EXPORT_TABLE_01": u1/******** directory=mydir
dumpfile=u1.dmp tables=emp
正在使用 BLOCKS 方法进行估计...
处理对象类型 TABLE_EXPORT/TABLE/TABLE_DATA
使用 BLOCKS 方法的总估计: 64 KB
```

```
处理对象类型 TABLE_EXPORT/TABLE/TABLE
. . 导出了 "U1"."EMP"                          5 KB        1 行
已成功加载/卸载了主表 "U1"."SYS_EXPORT_TABLE_01"
******************************************************************************
U1.SYS_EXPORT_TABLE_01 的转储文件集为:
   D:\EXP\U1.DMP
作业 "U1"."SYS_EXPORT_TABLE_01" 已于 08:55:46 成功完成
```

13.3.2 EXPDP 导出模式

EXPDP 支持多种导出模式，分别使用不同的参数来指定。

1. 全库导出模式

执行全库导出要求导出用户具有 DATAPUMP_EXP_FULL_DATABASE 角色，并指定参数 FULL。全库导出一般用于逻辑备份或数据库迁移。例:

```
expdp system/oracle directory = mydir dumpfile = full.dmp FULL = Y
```

2. Schema 导出模式

Schema 导出模式通过参数 SCHEMAS 指定，如果未指定任何导出模式，该导出模式为默认模式。如果用户具有 DATAPUMP_EXP_FULL_DATABASE 角色，可以导出任何 Schema，如果不具有 DATAPUMP_EXP_FULL_DATABASE 角色，则只能导出用户自己的 Schema。例:

```
expdp u1/u1 directory = mydir dumpfile = u1.dmp
expdp system/oracle directory = mydir dumpfile = schemas.dmp SCHEMAS = (u1,u2)
```

3. 表导出模式

表导出模式通过参数 TABLES 指定，可指定一个或多个表。例:

```
expdp u1/u1 directory = mydir dumpfile = u1.dmp TABLES = (dept,emp)
```

4. 表空间导出模式

表导出模式通过参数 TABLESPACES 指定，可指定一个或多个表空间。例:

```
expdp system/oracle directory = mydir dumpfile = ts.dmp TABLESPACES = (tbs1,tbs2)
```

5. 传输表空间模式

传输表空间模式通过参数 TRANSPORT_TABLESPACES 指定，只导出元数据，之后导入时需要将数据文件和导出文件复制到目标数据库。例:

```
expdp system/oracle directory = mydir dumpfile = ts.dmp
                  TRANSPORT_TABLESPACES = tbs1,tbs2
```

13.3.3 EXPDP 命令行参数

EXPDP 通过命令行参数控制导出行为，表 13-4 列出了常用的命令行参数。

表 13-4 EXPDP 命令行参数

序号	命令	说明
1	ATTACH	关联会话和导出任务
2	CLUSTER	指定是否使用集群资源
3	COMPRESSION	指定压缩项
4	CONTENT	指定导出内容
5	DIRECTORY	指定导出目录
6	DUMPFILE	指定导出文件名
7	ESTIMATE	指定空间估算方法
8	ESTIMATE_ONLY	指定是否只估算
9	EXCLUDE	指定不导出内容
10	FILESIZE	指定导出文件大小
11	FULL	指定全库导出
12	INCLUDE	指定导出内容
13	LOGFILE	指定导出日志文件
14	QUERY	指定过滤条件
15	SCHEMAS	指定导出的 Schema
16	TABLES	指定导出表
17	TABLESPACES	指定导出表空间
18	TRANSPORT_TABLESPACES	指定传输表空间

1. ATTACH

通过 ATTACH 可以连接当前会话和一个已存在的导出任务,进入任务的交互模式,这时可以查看导出任务状态、停止任务等。语法如下:

ATTACH [= [schema_name.] job_name]

如果当前存在多个导出任务,可指定用户名和任务名,任务名称可从视图 DBA_DATAPUMP_JOBS 中获得。如果当前用户只有一个导出任务,则不需指定任务名。比如用户 system 已经启动一个全库导出任务,开发另一个会话,看任务状态:

```
C:\> expdp system/oracle ATTACH
Export > status
作业: SYS_EXPORT_FULL_01
  操作: EXPORT
  模式: FULL
  状态: EXECUTING
处理的字节: 0
当前并行度: 1
作业错误计数: 0
转储文件: D:\EXP\FULL.DMP
  写入的字节: 4,096

Worker 1 状态:
  进程名: DW00
```

```
状态：EXECUTING
对象方案：SYSTEM
对象名：SYS_EXPORT_FULL_01
对象类型：DATABASE_EXPORT/SCHEMA/TABLE/TABLE_DATA
完成的对象数：68
Worker 并行度：1
```

2. CLUSTER

通过 CLUSTER 参数指定是否使用集群资源，默认是 YES，使用集群资源意味着集群中的多个实例都会启动进程参与导出。语法如下：

```
CLUSTER = [YES | NO]
```

例：

```
expdp system/oracle directory = mydir full = y CLUSTER = NO
```

3. COMPRESSION

通过 COMPRESSION 参数指定是否压缩导出数据，语法如下：

```
COMPRESSION = [ALL | DATA_ONLY | METADATA_ONLY | NONE]
```

ALL 指对元数据和表行数据都压缩，DATA_ONLY 指压缩表行数据，METADATA_ONLY 指压缩元数据，NONE 指不压缩，默认是 METADATA_ONLY。例：

```
expdp system/oracle directory = mydir full = y COMPRESSION = NONE
```

4. CONTENT

通过 CONTENT 参数指定导出内容，语法如下：

```
CONTENT = [ALL | DATA_ONLY | METADATA_ONLY ]
```

ALL 指导出元数据和表行数据，DATA_ONLY 指只导出表行数据，METADATA_ONLY 指只导出元数据，默认是 ALL。例：

```
expdp u1/u1 directory = mydir dumpfile = u1.dmp CONTENT = DATA_ONLY
```

5. DIRECTORY

通过 DIRECTORY 参数指定导出内容存储路径，语法如下：

```
DIRECTORY = directory_object
```

导出文件不能写磁盘路径，需要借助目录对象来完成，导出时用户需要具有对目录对象的读写权限，如果未指定目录对象，默认为 DATA_PUMP_DIR。例：

```
sqlplus system/oracle
SQL> create directory mydir as 'd:\exp';
SQL> grant read,write on directory mydir to u1;
expdp u1/u1 DIRECTORY = mydir dumpfile = u1.dmp
```

6. DUMPFILE

通过 DUMPFILE 参数指定导出文件名，默认为 expdat.dmp，语法如下：

```
DUMPFILE = [directory_object:]file_name[, … ]
```

如果指定参数 FILESIZE,则导出文件可能会有多个,可以在 DUMPFILE 参数中指定多个文件名。例:

```
DUMPFILE = full_1.dmp, full_2.dmp
```

如果文件数据量不定,可以指定%U,由系统自动生成文件名。例:

```
expdp system/oracle directory = mydir DUMPFILE = full%U full = y FILESIZE = 100M
```

7. ESTIMATE

通过 ESTIMATE 参数指定导出空间估算方法,语法如下:

```
ESTIMATE = [BLOCKS | STATISTICS ]
```

BLOCKS 指根据对象占用块数估算,STATISTICS 指根据对象的数据统计信息进行估算,STATISTICS 估算方法更为精确,但要求对象具有最新的统计信息。默认是 BLOCKS。

8. ESTIMATE_ONLY

通过 ESTIMATE_ONLY 参数指定是否只进行空间估算,语法如下:

```
ESTIMATE_ONLY = [YES | NO]
```

YES 指只估算不导出,默认是 NO。

```
C:\> expdp system/oracle ESTIMATE_ONLY = YES ESTIMATE = STATISTICS SCHEMAS = u1
Export: Release 11.2.0.1.0 - Production on 星期一 9 月 16 10:02:56 2013
Copyright (c) 1982, 2009, Oracle and/or its affiliates. All rights reserved.

连接到: Oracle Database 11g Enterprise Edition Release 11.2.0.1.0 - Production
With the Partitioning, OLAP, Data Mining and Real Application Testing options
自动启用 FLASHBACK 以保持数据库完整性.
启动 "SYSTEM"."SYS_EXPORT_SCHEMA_01": system/********
ESTIMATE_ONLY = YES ESTIMATE = STATISTICS SCHEMAS = u1
正在使用 STATISTICS 方法进行估计...
处理对象类型 SCHEMA_EXPORT/TABLE/TABLE_DATA
.  预计为 "U1"."T01"                                  5.433 KB
.  预计为 "U1"."T02"                                  5.433 KB
.  预计为 "U1"."EMP"                                  5.058 KB
.  预计为 "U1"."SYS_EXPORT_SCHEMA_01"                     0 KB
使用 STATISTICS 方法的总估计: 15.92 KB
作业 "SYSTEM"."SYS_EXPORT_SCHEMA_01" 已于 10:02:59 成功完成
```

9. EXCLUDE

通过 EXCLUDE 参数指定不执行导出的对象或对象类型,语法如下:

```
EXCLUDE = object_type[:name_clause] [, … ]
```

object_type 指定不导出对象的类型,具体类型参看视图 DATABASE_EXPORT_OBJECTS 中的 OBJECT_PATH 列,name_clause 指定过滤对象的表达式,包括操作符和

值。name_clause 语句必须使用双引号括起来,而双引号在命令行使用时需使用转义符
"\",否则会报语法错误,如果使用参数文件没有类似问题。EXCLUDE 语句可指定多
次。例:

```
expdp u1/u1 directory = mydir exclude = TABLE: \" = 'T01'\"
expdp system/oracle directory = mydir exclude = SCHEMA: \" = 'U1'\" full = y
```

或使用参数文件

```
par.txt
directory = mydir
exclude = SCHEMA: " = 'U1'"
full = y
expdp system/oracle parfile = par.txt
```

10. FILESIZE

通过 FILESIZE 参数指定导出文件大小,语法如下:

```
FILESIZE = integer [ B | KB | MB | GB | TB ]
```

例:

```
expdp system/oracle directory = mydir DUMPFILE = full%U full = y FILESIZE = 100M
```

11. FULL

通过 FULL 参数指定全库导出,语法如下:

```
FULL = [ YES | NO ]
```

12. INCLUDE

通过 INCLUDE 参数指定执行导出的对象或对象类型,语法如下:

```
INCLUDE = object_type [:name_clause] [, … ]
```

例:

```
INCLUDE = TABLE:\"IN ('EMPLOYEES', 'DEPARTMENTS')\"
INCLUDE = PROCEDURE
INCLUDE = INDEX:\"LIKE 'EMP%'\"
```

13. LOGFILE

通过 LOGFILE 参数指定导出日志文件名,语法如下:

```
LOGFILE = [directory_object:]file_name
```

例:

```
expdp u1/u1 directory = mydir dumpfile = u1.dmp LOGFILE = expdp.log
```

14. QUERY

通过 QUERY 参数指定表数据的过滤条件,语法如下:

```
QUERY = [schema.][table_name:] query_clause
```

query_clause 指定条件,就是 SELECT 语句中的 WHERE 语句内容。

```
QUERY = [schema.][table_name:] query_clause
query_clause 指定条件,就是 SELECT 语句中的 WHERE 语句内容。
C:\> expdp u1/u1 directory = mydir query = 'u1.T01:"WHERE id > 100"'
启动 "U1"."SYS_EXPORT_SCHEMA_01": u1/******** directory = mydir
    query = 'u1.T01:WHERE id > 100'
正在使用 BLOCKS 方法进行估计...
… …
.. 导出了 "U1"."T01"                         5.406 KB           1 行
.. 导出了 "U1"."T02"                         5.414 KB           2 行
已成功加载/卸载了主表 "U1"."SYS_EXPORT_SCHEMA_01"
******************************************************************
U1.SYS_EXPORT_SCHEMA_01 的转储文件集为:
  D:\EXP\EXPDAT.DMP
作业 "U1"."SYS_EXPORT_SCHEMA_01" 已于 14:45:44 成功完成
```

15. SCHEMAS

通过 SCHEMAS 参数指定导出的 Schema,语法如下:

```
SCHEMAS = schema_name [, …]
```

16. TABLES

通过 TABLES 参数指定导出的表,语法如下:

```
TABLES = [schema_name.]table_name[:partition_name] [, …]
```

17. TABLESPACES

通过 TABLESPACES 参数指定导出的表空间,语法如下:

```
TABLESPACES = tablespace_name [, …]
```

18. TRANSPORT_TABLESPACES

通过 TRANSPORT_TABLESPACES 参数指定导出的传输表空间,语法如下:

```
TRANSPORT_TABLESPACES = tablespace_name [, …]
```

```
表空间自包含检查:
SQL> EXECUTE dbms_tts.transport_set_check(ts_list => 'tbs1');
查询视图 transport_set_violations;

将表空间置只读
SQL> alter tablespace tbs1 read only;

执行导出
C:\> expdp system/oracle directory = mydir dumpfile = tbs1.dmp
TRANSPORT_TABLESPACES = tbs1
```

13.3.4 交互模式

进入交互模式的方法有两种：在导出命令执行过程中可按 Crtl+C 键进入交互模式，或在另一个终端执行带 ATTACH 参数的导出命令。交互模式下显示提示符 Export>，在该提示符下可执行一些命令，比如停止导出任务或增加导出文件，而当前的任务继续执行。表 13-5 列出了常用的交互模式命令。

表 13-5　EXPDP 交互模式命令

序号	命　　令	说　　明
1	ADD_FILE	增加导出文件
2	CONTINUE_CLIENT	退出交互模式
3	EXIT_CLIENT	退出导出会话，导出任务继续
4	FILESIZE	重新定义导出文件大小
5	HELP	显示可用命令
6	KILL_JOB	分离当前会话，终止当前任务
7	PARALLEL	增加或减少并行度
8	START_JOB	重新启动任务
9	STATUS	显示任务状态
10	STOP_JOB	停止当前任务

1. ADD_FILE

增加导出文件，语法如下：

```
ADD_FILE = [directory_object:] file_name [,…]
```

例：

```
Export> ADD_FILE = hr2.dmp, dpump_dir2:hr3.dmp
```

2. CONTINUE_CLIENT

当按 Ctrl+C 键进入交互模式后，可通过该命令退出交互模式，返回到原导出模式，日志信息依然会显示到终端窗口，语法如下：

```
CONTINUE_CLIENT
```

例：

```
Export> CONTINUE_CLIENT
```

3. EXIT_CLIENT

当按 Ctrl+C 键进入交互模式后，可通过该命令退出 Export，日志信息将不再显示到终端窗口，而任务将继续进行，语法如下：

```
EXIT_CLIENT
```

例：

```
Export> EXIT _CLIENT
```

4. FILESIZE

可以重新定义导出文件大小,语法如下:

FILESIZE = integer[B | KB | MB | GB | TB]

例:

Export > FILESIZE = 100MB

5. HELP

显示交互模式下可用命令,语法如下:

HELP

例:

Export > HELP

6. KILL_JOB

退出 Export,并删除导出任务,语法如下:

KILL_JOB

例:

Export > KILL_JOB

7. PARALLEL

增加或减少导出任务的并行度,语法如下:

PARALLEL = integer

例:

Export > PARALLEL = 10

8. START_JOB

在任务遇到异常停止或执行 STOP_JOB 命令后,可使用该命令重新启动导出任务,语法如下:

START_JOB

例:

Export > START_JOB

9. STATUS

显示任务的状态信息,语法如下:

STATUS[= integer]

使用 integer 指定间隔秒数,如果未指定只显示一次状态信息。

例:

Export > STATUS = 300

10. STOP_JOB

停止任务,语法如下:

STOP_JOB[= IMMEDIATE]

使用 IMMEDIATE 指立即停止任务,否则主进程会等待工作进程完成当前任务后停止。

例:

Export > STOP_JOB = IMMEDIATE

13.4 IMPDP

13.4.1 启动 IMPDP

Oracle 数据库软件安装后,在软件子目录 BIN 中包含 IMPDP 的可执行文件 impdp。如果是在 Windows 操作系统安装,该执行文件的路径已经添加到系统环境变量 PATH 中,可通过系统命令 SET 或 ECHO %PATH% 查看。如果是 Linux 系统,则该执行文件路径需要用户手动添加到系统环境变量 PATH 中,可通过系统命令 ENV 或 ECHO $PATH 查看。

打开命令行窗口,在命令提示符下输入命令"impdp - help",显示命令 impdp 的用法,如图 13-8 所示。

图 13-8 命令 impdp 用法

IMPDP 命令的语法如下:

impdp username/password[@connect_identifier] keyword = value [,keyword = value, …]

其中,username/password 指定执行导入任务的用户名和口令,connect_identifier 指定连接远程数据库的网络服务名,keyword 指定命令参数,常用的有 directory、dumpfile 等,directory 指定一个目录对象,dumpfile 指定导入文件名称。下例演示将 u1 用户的导出信息导入给 u2 用户。

```
C:\> expdp u1/u1 directory = mydir dumpfile = u1.dmp
. . 导出了 "U1"."EMP"                    5 KB         1 行
. . 导出了 "U1"."T01"                    5.421 KB     3 行
. . 导出了 "U1"."T02"                    5.421 KB     3 行
…

C:\> impdp u2/u2 directory = mydir dumpfile = u1.dmp
    remap_schema = u1:u2
…
启动 "U2"."SYS_IMPORT_FULL_01": u2/ ******** directory = mydir
dumpfile = u1.dmp remap_schema = u1:u2
…
. . 导入了 "U2"."EMP"                    5 KB         1 行
. . 导入了 "U2"."T01"                    5.421 KB     3 行
. . 导入了 "U2"."T02"                    5.421 KB     3 行
…
作业 "U2"."SYS_IMPORT_FULL_01" 已经完成.
```

13.4.2 IMPDP 导入模式

IMPDP 支持多种导入模式，分别使用不同的参数来指定。

1. 全库导入模式

执行全库导入要求导入用户具有 DATAPUMP_IMP_FULL_DATABASE 角色，并指定参数 FULL。全库导入一般用于逻辑备份或数据库迁移。例：

```
impdp system/oracle directory = mydir dumpfile = full.dmp FULL = Y
```

2. Schema 导入模式

Schema 导入模式通过参数 SCHEMAS 指定，导入的源文件可以是任何模式的导出文件，如果用户具有 DATAPUMP_IMP_FULL_DATABASE 角色，可以导入任何 Schema。例：

```
impdp system/oracle directory = mydir dumpfile = full.dmp schemas = u1
```

3. 表导入模式

表导入模式通过参数 TABLES 指定，可指定一个或多个表。例：

```
impdp u1/u1 directory = mydir dumpfile = u1.dmp TABLES = t01
```

4. 表空间导入模式

表空间导入模式通过参数 TABLESPACES 指定，可指定一个或多个表空间。例：

```
impdp system/oracle directory = mydir dumpfile = full.dmp TABLESPACES = tbs
```

5. 传输表空间模式

传输表空间的元数据可通过 EXPDP 导出，也可以通过 NETWORK_LINK 直接从远

程数据库获取,实际表空间的数据文件通过 TRANSPORT_DATAFILES 指定。例:

```
impdp system/oracle directory = mydir dumpfile = transport.dmp
    TRANSPORT_DATAFILES = /u01/app/tbs.dbf
impdp system/oracle directory = mydir NETWORK_LINK = link1
    TRANSPORT_TABLESPACES = tbs TRANSPORT_DATAFILES = /u01/app/tbs.dbf
```

13.4.3 IMPDP 命令行参数

IMPDP 通过命令行参数控制导出行为,表 13-6 列出了常用的命令行参数。

表 13-6 IMPDP 命令行参数

序号	命令	说明
1	ATTACH	关联会话和导入任务
2	CLUSTER	指定是否使用集群资源
3	CONTENT	指定导入内容
4	DIRECTORY	指定导入目录
5	DUMPFILE	指定导入文件名
6	EXCLUDE	指定不导入内容
7	FULL	指定全库导入
8	INCLUDE	指定导入内容
9	LOGFILE	指定导入日志文件
10	NETWORK_LINK	指定数据库链接
11	QUERY	指定过滤条件
12	REMAP_SCHEMA	指定 Schema 映射关系
13	REMAP_TABLE	指定表名映射关系
14	REMAP_TABLESPACE	指定表空间名映射关系
15	SCHEMAS	指定导入的 Schema
16	TABLE_EXISTS_ACTION	指定表存在情况下的处理方式
17	TABLES	指定导入表
18	TABLESPACES	指定导入表空间
19	TRANSPORT_DATAFILES	指定传输表空间文件名
20	TRANSPORT_TABLESPACES	指定传输表空间

1. ATTACH

通过 ATTACH 可以连接当前会话和一个已存在的导入任务,进入任务的交互模式,这时可以查看导入任务状态、停止任务等。语法如下:

```
ATTACH [ = [schema_name.] job_name]
```

如果当前存在多个导入任务,可指定用户名和任务名,如果当前用户只有一个导出任务,则不需指定任务名。比如用户 system 已经启动一个全库导入任务,开启另一个会话,看任务状态。例:

```
impdp system/oracle ATTACH
```

2. CLUSTER

通过 CLUSTER 参数指定是否使用集群资源,默认是 YES,使用集群资源意味着集群中的多个实例都会启动进程参与导入。语法如下：

```
CLUSTER = [YES | NO]
```

例：

```
impdp system/oracle directory = mydir dumpfile = full.dmp full = y CLUSTER = NO
```

3. CONTENT

通过 CONTENT 参数指定导入内容，语法如下：

```
CONTENT = [ALL | DATA_ONLY | METADATA_ONLY]
```

ALL 指导入元数据和表行数据；DATA_ONLY 指只导入表行数据，不创建表；METADATA_ONLY 指只导入元数据，默认是 ALL。例：

```
impdp u1/u1 directory = mydir dumpfile = u1.dmp CONTENT = DATA_ONLY
```

4. DIRECTORY

通过 DIRECTORY 参数指定导入内容存储路径，语法如下：

```
DIRECTORY = directory_object
```

例：

```
impdp u1/u1 DIRECTORY = mydir dumpfile = u1.dmp
```

5. DUMPFILE

通过 DUMPFILE 参数指定导入文件名，默认为 expdat.dmp，语法如下：

```
DUMPFILE = [directory_object:]file_name[, …]
```

如果导入文件是多个可分别列出，例：DUMPFILE=full_1.dmp，full_2.dmp，也可以指定%U，系统自动检查符合规则的文件。

6. EXCLUDE

通过 EXCLUDE 参数指定不执行导入的对象或对象类型，语法如下：

```
EXCLUDE = object_type[:name_clause][, …]
```

例：

```
impdp u1/u1 directory = mydir dumpfile = u1.dmp EXCLUDE = TABLE: \" = 'T01'\"
```

7. FULL

通过 FULL 参数指定全库导入，语法如下：

```
FULL = [YES | NO]
```

例：

```
impdp system/oracle directory = mydir dumpfile = full.dmp FULL = Y
```

8. INCLUDE

通过 INCLUDE 参数指定执行导入的对象或对象类型，语法如下：

```
INCLUDE = object_type [:name_clause] [, …]
```

例：

```
INCLUDE = TABLE:\"IN ('EMPLOYEES', 'DEPARTMENTS')\"
INCLUDE = PROCEDURE
INCLUDE = INDEX:\"LIKE 'EMP%'\"
```

9. LOGFILE

通过 LOGFILE 参数指定导入日志文件名，语法如下：

```
LOGFILE = [directory_object:]file_name
```

例：

```
impdp u1/u1 directory = mydir dumpfile = u1.dmp LOGFILE = expdp.log
```

10. NETWORK_LINK

通过 NETWORK_LINK 参数指定数据库间的网络连接，通过网络直接导入，语法如下：

```
NETWORK_LINK = source_database_link
```

例：

```
impdp system/oracle NETWORK_LINE = link1 schemas = u1
```

11. QUERY

通过 QUERY 参数指定表数据的过滤条件，语法如下：

```
QUERY = [schema.][table_name:] query_clause
```

query_clause 指定条件，就是 SELECT 语句中的 WHERE 语句内容。例：

```
impdp u1/u1 directory = mydir dumpfile = u1.dmp query = 'u1.T01:"WHERE id>100"'
```

12. REMAP_SCHEMA

通过 REMAP_SCHEMA 参数将源 Schema 的数据导入到目标 Schema，语法如下：

```
REMAP_SCHEMA = source_schema:target_schema
```

例：

```
impdp u2/u2 directory = mydir dumpfile = u1.dmp REMAP_SCHEMA = u1:u2
```

13. REMAP_TABLE

通过 REMAP_TABLE 参数可以在导入过程中修改表名，语法如下：

```
REMAP_TABLE = [schema.]old_tablename[.partition]:new_tablename
```

例：

```
impdp u1/u1 directory = mydir dumpfile = u1.dmp REMAP_TABLE = emp1:e1
```

14. REMAP_TABLESPACE

通过 REMAP_TABLESPACE 参数将源表空间的内容存储到目标表空间，语法如下：

```
REMAP_TABLESPACE = source_tablespace:target_tablespace
```

例：

```
impdp u1/u1 directory = mydir dumpfile = u1.dmp REMAP_TABLESPACE = users:tbs
```

15. SCHEMAS

通过 SCHEMAS 参数指定导入的 Schma，语法如下：

```
SCHEMAS = schema_name [, …]
```

16. TABLE_EXISTS_ACTION

通过 TABLE_EXISTS_ACTION 参数指定导入时如果表已经存在的处理方式，语法如下：

```
TABLE_EXISTS_ACTION = [SKIP | APPEND | TRUNCATE | REPLACE]
```

SKIP 指对表不做任务处理，APPEND 将以追加的方式导入数据，TRUNCATE 指先删除表中数据再导入数据，REPLACE 指删除表然后重新创建并导入数据。默认是 SKIP。

17. TABLES

通过 TABLES 参数指定导入的表，语法如下：

```
TABLES = [schema_name.]table_name[:partition_name] [, …]
```

18. TABLESPACES

通过 TABLESPACES 参数指定导入的表空间，语法如下：

```
TABLESPACES = tablespace_name [, …]
```

19. TRANSPORT_DATAFILES

通过 TRANSPORT_DATAFILES 参数指定传输表空间导入的数据文件名，语法如下：

```
TRANSPORT_DATAFILES = datafile_name
```

例：

```
impdp system/oracle directory = mydir dumpfile = transport.dmp
    TRANSPORT_DATAFILES = /u01/app/tbs1.dbf
```

20. TRANSPORT_TABLESPACES

通过 TRANSPORT_TABLESPACES 参数指定导入的传输表空间，同时需要指定 NETWORK_LINK 来获取元数据，语法如下：

```
TRANSPORT_TABLESPACES = tablespace_name [, …]
```

例：

```
impdp system/oracle directory = mydir NETWORK_LINK = link1
```

TRANSPORT_TABLESPACES = tbs TRANSPORT_DATAFILES = /u01/app/tbs.dbf

13.4.4 交互模式

进入交互模式的方法有两种：在导入命令执行过程中可按 Crtl+C 键进入交互模式，或在另一个终端执行带 ATTACH 参数的导入命令。交互模式下显示提示符 Import>，在该提示符下可执行一些命令。表 13-7 列出了常用的交互模式命令。

表 13-7　IMPDP 交互模式命令

序号	命　　令	说　　明
1	CONTINUE_CLIENT	退出交互模式
2	EXIT_CLIENT	退出导出会话，导出任务继续
3	HELP	显示可用命令
4	KILL_JOB	分离当前会话，终止当前任务
5	PARALLEL	增加或减少并行度
6	START_JOB	重新启动任务
7	STATUS	显示任务状态
8	STOP_JOB	停止当前任务

1. CONTINUE_CLIENT

当按 Ctrl+C 键进入交互模式后，可通过该命令退出交互模式，返回到原导入模式，日志信息依然会显示到终端窗口，语法如下：

CONTINUE_CLIENT

例：

Import > CONTINUE_CLIENT

2. EXIT_CLIENT

当按 Ctrl+C 键进入交互模式后，可通过该命令退出 Import，日志信息将不再显示到终端窗口，而任务将继续进行，语法如下：

EXIT_CLIENT

例：

Import > EXIT _CLIENT

3. HELP

显示交互模式下可用命令，语法如下：

HELP

例：

Import > HELP

4. KILL_JOB

退出 Import,并删除导出任务,语法如下:

```
KILL_JOB
```

例:

```
Import > KILL_JOB
```

5. PARALLEL

增加或减少导入任务的并行度,语法如下:

```
PARALLEL = integer
```

例:

```
Import > PARALLEL = 10
```

6. START_JOB

在任务遇到异常停止或执行 STOP_JOB 命令后,可使用该命令重新启动导入任务,语法如下:

```
START_JOB
```

例:

```
Import > START_JOB
```

7. STATUS

显示任务的状态信息,语法如下:

```
STATUS[ = integer]
```

使用 integer 指定间隔秒数,如果未指定只显示一次状态信息。例:

```
Import > STATUS = 300
```

8. STOP_JOB

停止任务,语法如下:

```
STOP_JOB[ = IMMEDIATE]
```

使用 IMMEDIATE 指立即停止任务,否则主进程会等待工作进程完成当前任务后停止。例:

```
Import > STOP_JOB = IMMEDIATE
```

小　　结

本章讲解了 Oracle 数据库的常用工具使用,包括 SQL * Plus、SQL * Loader、EXPDP 和 IMPDP。SQL * Plus 是 Oracle 数据库的客户端工具,稳定且功能很强;SQL * Loader

是 Oracle 数据库的数据导入工具,可将文本数据导入到数据库表中;EXPDP 和 IMPDP 是 Oracle 数据库的数据移植工具,可完成不同平台间的数据迁移。

思 考 题

1. 使用 sqlldr 将如下数据导入到表 stu 中,写出所有操作脚本。

表 stu: create table stu(id number, name varchar2(20), gender varchar2(20), birth date)

需导入数据:

data.txt

1,张三,男,1980-06-13
2,李四,女,1981-10-22

2. 使用 sqlldr 将如下数据导入到表 stu(参见思考题 1 中 stu 表)中,写出所有操作脚本。

需导入数据:

data.txt

1,张三,男,34,1980-06-13
2,李四,女,33,1981-10-22

3. 使用 expdp 可以完成哪些级别的数据导出?

4. 创建表空间 tbs1,创建用户 u1,u1 的默认表空间是 tbs1;
创建表空间 tbs2,创建用户 u2,u2 的默认表空间是 tbs2;
授予用户 u1 和 u2 两个角色:connect、resource 角色;
以 u1 用户登录,创建表 t1(id number),插入一行数据;
导出 u1 用户的所有 Schema 对象;
将导出的 u1 用户的信息导入到用户 u2 中。